Ethical Economy

Studies in Economic Ethics and Philosophy

Volume 72

Ethical Economy describes the theory of the ethical preconditions of the economy and of business as well as the theory of the ethical foundations of economic systems. It analyzes the impact of rules, virtues, and goods or values on economic action and management. *Ethical Economy* understands ethics as a means to increase trust and to reduce transaction costs. It forms a foundational theory for business ethics and business culture. The Series *Ethical Economy. Studies in Economic Ethics and Philosophy* is devoted to the investigation of interdisciplinary issues concerning economics, management, ethics, and philosophy. These issues fall in the categories of economic ethics, business ethics, management theory, economic culture, and economic philosophy, the latter including the epistemology and ontology of economics. Economic culture comprises cultural and hermeneutic studies of the economy. One goal of the series is to extend the discussion of the philosophical, ethical, and cultural foundations of economics and economic systems. The series is intended to serve as an international forum for scholarly publications, such as monographs, conference proceedings, and collections of essays. Primary emphasis is placed on originality, clarity, and interdisciplinary synthesis of elements from economics, management theory, ethics, and philosophy.

The book series has been accepted into SCOPUS (March 2019) and will be visible on the Scopus website within a few months.

Javier Pinto-Garay · Germán Scalzo

Justice and Corporate Excellence

A Common Good Approach to Business and Professional Ethics

Springer

Javier Pinto-Garay
Business Ethics and HR Management
Universidad de los Andes
Santiago, Chile

Germán Scalzo
Business Ethics
Universidad Panamericana
Mexico City, Mexico

ISSN 2211-2707 ISSN 2211-2723 (electronic)
Ethical Economy
ISBN 978-3-032-12188-2 ISBN 978-3-032-12189-9 (eBook)
https://doi.org/10.1007/978-3-032-12189-9

This Springer imprint is published by the registered company Springer Nature Switzerland AG
The registered company address is: Gewerbestrasse 11, 6330 Cham, Switzerland

Foreword

Ethics often presents itself in paradoxes. It resists simple definitions and thrives in contrasts. To understand it, one must accept its dual nature: at once ancient and ever new, slow and fast, creative and pragmatic. This is why any attempt to approach ethics requires both patience and imagination, reflection and practice.

Ethics is both old and young at the same time. For more than twenty-six centuries, Western thought has asked about good and evil, about what is right and what is wrong. Since the time when Thales of Miletus fell into a well while gazing at the stars, philosophers discovered that more important than asking about the origin of the world is asking about who the human being is.

Over these centuries, we have debated endlessly, though we have reached few definitive agreements. Human action always has something elusive about it—something that escapes us, and it is precisely this elusiveness that belongs to ethics. That is why consensus is difficult: ethics often slips away like water through our fingers. And yet, the plurality of traditions, arguments, and perspectives has woven a rich tapestry, one that reveals the complexity and depth of ethics itself.

Ethics is old, yes, but it is also young. It carries the vitality of a life that constantly reinvents itself. Human life is a sequence of questions and answers, problems and solutions, which in turn generate new challenges. In this ongoing symphony, ethics plays its role. Human progress solves problems, but at the same time creates new ones. That is why ethics is like the phoenix, always rising from its ashes, ready to confront the moral dilemmas that every new stage of life brings.

Ethics is both slow and fast. It needs time, reflection, and prudence. It resists haste. If forced into quick responses, it becomes calculating ethics (weighing costs and benefits) or rule-based ethics (mechanical application of norms). But the question that truly matters to ethics—what kind of person am I becoming through my actions?—cannot be answered mechanically. It requires craftsmanship, shaping nothing less than one's own life. And this demands time.

Yet ethics is also quick. It has the speed of an intuition formed by experience. Just as when learning to drive one must first consciously think about every movement, until a moment comes when the motions are second nature, so too ethics forms habits. Over time, moral habits make doing good natural and spontaneous. They become like the reed that bends with the wind without breaking. Moral habits help us adapt to circumstances while remaining faithful to our deepest principles.

Ethics is creative and pragmatic at once. It calls us to do good, but acknowledges that the good can be pursued in many different ways. It directs us toward the goal we must seek—happiness understood not as passive pleasure but as fullness of life achieved through effort and growth. Ethics points to the goal, but does not dictate the path: this is entrusted to our freedom and responsibility. And so, many roads open before us.

Ethics is creative because it does not have only one answer. Each situation is unique, and so each response is different; each person is unique, and there are as many responses as there are people. Ethics is always conjugated in the first person. Whenever we try to reduce it to formulas or stereotypes, we lose the wisdom of practical reasoning—knowing what is best to do here and now.

But this creativity does not mean idealism or naïveté. Ethics is realistic, because decisions are always made *here and now*, within real conditions that cannot be ignored. Ethics is subjective at the level of prudential judgment, though not at the level of principles. It is realistic, because people and things are what they are, and they tell us how they must be treated. We cannot treat them as we please.

The path of ethics is full of paradoxes, and it must be walked carefully. Javier Pinto and Germán Scalzo offer in this book a reliable guide for that journey into the world of business and professional ethics. It is a book that is fresh and creative, yet firmly rooted in a millenary tradition, and one that helps us reflect on today's managerial challenges without falling into simplistic answers.

One of the book's distinctive contributions is its systematic framework for business ethics, inspired by the Aristotelian and Thomistic tradition. Too often, business ethics is taught or practiced in a fragmented way. Here, however, we find a coherent architecture: business ethics as a theory of justice and the common good, applied across three domains of corporate life—organizational, commercial, and public. These correspond to Aristotle's threefold division of justice: distributive, commutative, and legal-political.

The book also insists on the integration of principles and practices. Principles such as justice, dignity, the Golden Rule, or care for the environment give direction to action; practices such as lean management, ESG policies, or even collusion illustrate how principles are embodied—or violated—in reality. Only by holding both dimensions together can business ethics avoid becoming either empty theory or mere pragmatism.

Equally valuable is the book's proposal of a neo-Aristotelian stakeholder theory. Beyond Friedman's minimalist view and Freeman's pluralist approach, Pinto and Scalzo ground stakeholder relationships in justice. They show that distributive justice guides the firm's internal life, recognizing contribution and merit; commutative justice shapes fair exchanges in the market, based on mutual benefit; and legal-political justice frames the firm's responsibilities toward society, law, and the common good.

Finally, the authors distinguish between minimum and maximum justice. Minimum justice means compliance with laws and customs; maximum justice calls for responsibility, generosity, and the "logic of gift." Business, then, is not only about avoiding harm but about actively promoting the flourishing of others—workers, consumers, and communities.

This is a book that is systematic yet accessible, philosophical yet practical, rigorous yet inspiring. It invites us to see business not only as production (*poiesis*) but also as action (*praxis*); not only as efficiency but also as prudence; not only as self-interest but also as friendship and contribution to the common good.

For Pinto and Scalzo, excellence in business is inseparable from justice. Justice provides the essential measure for corporate practices, ordering relationships inside the firm, in the market, and with society. Yet excellence goes a step further: it is not only about giving each their due, but about doing so with virtue, responsibility, and generosity. Excellence makes justice "excel," transforming it from a minimum requirement into a maximum aspiration oriented toward the common good.

The reader will find here more than a manual: it is an invitation to live business ethics as a path to excellence, and to discover that in doing good, one can also discover the true joy of flourishing.

Joan Fontrodona
Business Ethics Professor, CaixaBank
Chair on Sustainability and Social
Impact. IESE Business
School-University of Navarra
Pamplona, Spain

Competing Interests The authors have no competing interests to declare that are relevant to the content of this manuscript.

Contents

About the Authors

Javier Pinto-Garay is an Associate Professor of Business Ethics at the School of Business and Economics, Universidad de los Andes, Chile. He has published several articles, chapters, and books on Business and Professional Ethics, and Philosophy of Work. He is currently working on virtue ethics and new Natural Law applied to business practices and work. He holds a B.S. in Philosophy (Universidad de los Andes), an M.Sc. in Organizational Behavior and HR Management (Pontificia Universidad Católica de Chile) and a Ph.D. in Philosophy (University of Navarra).

Germán Scalzo is a tenured Full Professor of Business Ethics at the Universidad Panamericana's Faculty of Economic and Business, in Mexico City. He holds an International Ph.D. in Philosophy and a Master's degree in Governance and Culture of Organizations from the University of Navarra, Spain, in addition to a bachelor's degree in business from Austral University, Argentina. He has published numerous articles, chapters and books on economics and business through an anthropological and philosophical lens. His current research includes virtue ethics and the logic of gift.

Abbreviations

In Eth	Tomas Aquinas' Commentary on Aristotle's Ethics
In Pol	Tomas Aquinas' Commentary on Aristotle's Politics
NE	Aristotle's Nicomachean Ethics
Pol	Aristotle's Politics
Rhet	Aristotle's Rhetoric
S.Th.	Tomas Aquinas' Summa Theologica
SCG	Tomas Aquinas' Summa Contra Gentes
WN	Adam Smith's An Inquiry into the Nature and Causes of the Wealth of Nation

Chapter 1
Introduction

Abstract This introductory chapter explains the motivation behind the book and outlines its epistemological and pedagogical foundations. In response to the lack of a coherent panoramic view of business ethics, it provides a systematic integration of the field's main ideas, concepts, current debates, and philosophical arguments, offering a structured framework for both academic teaching and practical application. It addresses the moral principles and practices that encompass the most relevant issues and problems of business ethics in a systematic manner, guided by a neo-Aristotelian and New Natural Law perspective that understands business ethics as a theory of justice and the common good applied to work, commerce, and society. Building on this conceptual distinction, the book establishes a threefold division of practices—organizational, commercial, and corporate—mirroring the Aristotelian account of distributive, commutative, and public justice. By articulating this preliminary framework, this book lays the foundation for a systematic and comprehensive analysis of business ethics in terms of excellence, grounded in personal virtue and the common good.

Keywords Business ethics · Neo-Aristotelianism · New natural law · Justice · Common good · Principles · Practices · Virtue ethics · Organizational ethics · Commercial ethics · Corporate ethics

We were originally motivated to write this book by what we perceived as the need for a more systematic organization of the topics and areas associated with business ethics. In our academic work, we found that, in the absence of such structure, the field was difficult to teach, not only in undergraduate, but also in graduate, and MBA programs—the latter, in fact, presenting an additional challenge, as the case method does not leave much space for explaining theory. Without a systematic integration of the many ideas, concepts, current debates, opinions, and philosophical arguments involved, it was very hard to provide a coherent panoramic view of this discipline in class.

As a result, and after a few years of work, we have now arrived at a proposal of principles and practices that encompasses business ethics' most relevant issues and

J. Pinto-Garay and G. Scalzo, *Justice and Corporate Excellence*, Ethical Economy 72,
https://doi.org/10.1007/978-3-032-12189-9_1

problems, approaching them in a systematic manner derived from a specific moral philosophy tradition: neo-Aristotelianism and New Natural Law.

Indeed, based on this epistemological foundation, our central aim of providing a systematic organization of business ethics theory is inspired not only by pedagogical ends, but also by what we may call epistemological reasons. In our view, business ethics, grounded in a neo-Aristotelian tradition, is ultimately a theory of justice and the common good, applied comprehensively to work, commerce, and society. Accordingly, in developing this neo-Aristotelian approach to business ethics, the activity of the firm should be understood as structured around three principal areas—organizational, commercial, and public—mirroring the threefold division of justice in neo-Aristotelian thought: distributive, commutative, and public (Scalzo et al. 2021). It is within the logic of this threefold framework that we devoted ourselves to the task of explaining business activities in terms of excellence; the ethical principle understood as the attainment of the highest possible standards of business practice in accordance with personal virtue and common good.

1.1 Principles and Practices: A Preliminary Distinction

Perhaps one of the most compelling articulations of business ethics in neo-Aristotelian terms—particularly in terms of excellence—comes from MacIntyre (2016, pp. 170–171) when he explains that:

> The common goods of those at work together are achieved in producing goods and services that contribute to the life of the community and in becoming excellent at producing them.…When, instead, teams of workers cooperated in taking each car through the different stages of its production, taking responsibility as a team for the quality of the end product, things went much better both for the cars and for the workers. The ends informing the workers' activity are now those of achieving through shared deliberation and decision the making of an excellent car and of becoming excellent in making such cars. It matters that they understand what they are doing and that their standards are ones that they have made their own, not standards imposed by external managerial control. They share direction toward a common good.

This example, in fact, provides a description of good work and industry based on the essential features of a virtue ethics approach to business and market practices, namely, the principles of common good and excellence. On the other hand, what MacIntyre describes philosophically is recognizable in the integration of Participatory Management practices like Quality Circles or Lean Manufacturing. These specific methodologies of work are, in general, recognized as ethically good, particularly when implemented for the sake of improving the quality of the human experience of work along with increased productivity. Indeed, we might consider, for instance, that Ponzi schemes, collusion, implementation of ESG policies, or child labor are examples of good or bad practices in ethical terms. On the other hand, the common good, justice, the dignity of the worker, the golden rule, or caring for the environment are moral principles.

Now, MacIntyre's description provides a good view of what a business ethics theory should consider epistemologically as the need to distinguish between principles and practices as two dimensions to be considered integrally in a more complete and objective ethical analysis of business and work activities, rather than as two totally separate realms of reflection. In other words, a complete analysis of business activity in ethical terms must integrate both practices and principles. Taking into account such integration, there are two preliminary (and quite significant) considerations to keep in mind for understanding how this book is organized. The first one is the importance of understanding practices without limiting business and professional ethics theory to a mere matter of moral principles, and vice versa. The second consideration entails what we mean by 'practices' and how they can be defined.

As for the first consideration, it is clear that moral principles are important for assessing all business and professional ethics practices. Discussing said practices without a deeper and grounding reflection on the moral principles behind them is a merely descriptive exercise—one that may at times present contradictory examples that can be seen, for instance, when comparing codes of conduct with different and sometimes contradictory definitions of the same practice. In other words, moral principles provide a conceptual framework needed to understand why certain definitions of practices might be conceptually incorrect and, more importantly, in practical terms.

In a similar manner, Finnis (2011) explains that conceptual frameworks—i.e., basic goods and moral principles—are essential in the practical reflection and decision-making processes of those who participate in the various spheres of human life (such as judges, rulers, and even citizens). According to him, to be fully reasonable, one must remain open to every basic aspect of human flourishing, to every basic form of human good. In this sense, moral principles are the practical criteria that constitute a form of practical reasonableness precisely because they are directed toward one or more requirements of human flourishing—in other words, they are directed toward the basic forms of human good, such as beauty, truth, or friendship. The Golden Rule—i.e., treat others as you would like to be treated—is, for instance, a primary moral principle directed toward a basic form of human good, namely, friendship (Finnis 2011).

But, in the other direction, a reflection on moral principles that seeks to guide business and professional activity without any knowledge and consideration of the various business and professional practices, becomes—paraphrasing Finnis (2011)—a mere formalism, that is, one that can hardly contribute to practical reasonableness. For instance, a reflection based solely on a well-developed concept of justice cannot provide meaningful guidance when considering whether an activity is reasonable to carry out if the person making the decision does not also take into account what that activity is and what its scope entails. In this sense, a substantive understanding of what the practices specific to a given domain are provides precisely the knowledge of the scope of an activity and thus of its proximate end. In this vein, Aquinas explains in his commentary on Aristotle's ethics (Aquinas 1961), that for a practical science whose end is action, we must know the activities or operations by which a determined effect follows from a determined cause. These activities and operations are what we call practices.

Accordingly, contributing to reasonable decision-making from a conceptual framework is not truly feasible unless, in addition to a robust consideration of moral principles, the decision-maker also has real knowledge of the wide range of possibilities and opportunities, inclinations and capabilities, and everything that, in the realm of human activity, is known descriptively as practices. Indeed, one whose knowledge of the concrete reality of human affairs is limited or nonexistent is ultimately someone who will struggle to accurately judge the practical implications of moral principles. In this sense, there is a mutual (though not symmetrical) dependence between the description of human affairs and the moral consideration of action in terms of its principles (Finnis 2011).

As for the second consideration (i.e., what do we mean by 'practices' and how can they be defined), practices refer to a material and formal description of a kind of acting. According to a virtue ethics approach, a practice can be categorized following the object of an action. According to Rhonheimer (2005), the object of an action shows '*what* one does when one does something'. In other words, 'what one does' characterizes the related practice and allows us to classify different actions under the same definition.

Now, the definition of practices in terms of objects is achieved by considering both the material and formal dimensions of the action. Indeed, this material-formal consideration in defining practices is precisely what Thomas Aquinas would regard as the description of an action in terms of its object—that is, without the need to take into account either the particular circumstances in which these actions are performed or the ultimate end of the person performing them (Finnis 2005).

According to Sousa-Lara (2008), for Thomas Aquinas, the material dimension—or, more precisely, the almost (quasi) material dimension—is that which is revealed as 'what is sought' or 'what one intends to do' (*res volita*) materially or in executive terms. That is, in the case of the practice of collusion, communicating information related to one's own productive or commercial activity to competitors in order to influence prices outside regular market activities. Moreover, the material dimension also incorporates material goods that constitute what we can call an external thing (*res extensa*) of the object of the action which, as John of St. Thomas (1964) explains, is not strictly the operation materially carried out, but rather what is operated in the action—that is, the content of the object. These are, following the same example of collusion, 'the written messages with commercial information exchanged between executives who belong to competing companies.'

In contrast, the formal dimension of the object of the action—its proximate end—is defined according to the logic of a practical benefit (*ratio boni*), without which the practice would not be desired or pursued by those who carry it out. This practical benefit, proper to a proximate end, is what is specifically regarded as suitable by those engaging in the practice and, more importantly, centrally defines a practice. Thus, in the example of collusion, the proximate end would be 'to prevent the entry of direct competition into the market by means unrelated to those proper to a free market.' Here, collusion has an intrinsically evil formal structure. This is not due to the fact that ‘some information was shared between competitors’ (which is the quasi-materiality of the object), nor to the desire to maintain or improve the company’s economic income (i.e., a rather a remote end of the agent—*finis operantis*—which is, in this case, totally legitimate). Rather, it is evil because, while done with a view to a suitable proximate end, it ultimately goes against a moral principle—in this case, justice and the common good—necessarily restricting the legitimate participation of other individuals who could contribute to and benefit from market activities and, therefore, the common good of a community.

Therefore, even though the quasi-material dimension and the content of the object are highly relevant in identifying a practice, the materiality of the object is not essentially determinative in specifying a practice. For example, we can observe that two opposing practices, such as theft and restitution, may share the same material content—namely, money (Salmanticenses 1878)—or, in the example of collusion, ‘communicating with competitors on matters related to one’s own productive or commercial activity’ could constitute the same quasi-material dimension in both opposing practices such as collusion—which is ethically bad—and clustering—which is considered good. In this sense, it becomes clear that the specificity of a type of practice derives not from its materiality, but from its formality and, specifically, its proximate end (Aquinas 1947, *S.Th.*, I-II, q.18. a.5, ad.2).

Thus, structuring a text on business and professional ethics involves considering that proximate ends not only serve to evaluate the morality of a given practice but also enable the grouping of various practices into specific species. For instance, monopolies, cartels, and collusive agreements can be classified as practices of the same kind—namely, those related to the functioning of the free market—whereas pyramid schemes and predatory lending are more appropriately categorized under practices associated with commercialization. Moreover, the criterion by which proximate ends can be grouped into the same species of practices is determined by the nature of the relationships in which those proximate ends are embedded. These relationships, in turn, are defined in virtue ethics by the moral principle of justice and, particularly, according to the three parts of justice that we will explain further on. On this basis, practices in business and professional ethics are classified here into three distinct species, namely organizational practices, commercial practices, and public practices.

References

Aquinas, T. (1274/1947). *Summa theologiae* (fathers of the English Dominican Province, Trans.). New York: Benziger Brothers

Aquinas, T. (1271/1961). *Commentary on Aristotle’s Nicomachean ethics*. New York: Dumb Ox Books.

Finnis, J. (2005). *Aquinas’ moral, political, and legal philosophy*. https://plato.stanford.edu/entries/aquinas-moral-political/. Accessed 22 Aug 2025.

Finnis, J. 2011. *Human rights and common good: Collected essays*. Vol. 3. Oxford: Oxford University Press.

John of St. Thomas. 1964. *Cursus theologicus*. Vol. 5. Paris: Desclée & Cie.
MacIntyre, A. 2016. *Ethics in the conflicts of modernity: An essay on desire, practical reasoning, and narrative*. Cambridge: Cambridge University Press.
Rhonheimer, M. 2005. La prospettiva della persona agente e la natura della ragione pratica. In *Camminare nella Luce. Prospettive della teologia morale a 10 anni da Veritatis splendor*, ed. L. Melina and J. Noriega, 169–224. Rome: Lateran University Press.
Salmanticenses. 1878. *Cursus theologicus*. Vol. 5. Paris: V. Palmé–J. Albanel.
Scalzo, G., J. Pinto-Garay, and K. Akrivou. 2021. CSR and virtue ethics: the common good of firms, markets, and civil society. In *The Routledge companion to corporate social responsibility*, 78–87. London: Routledge. https://doi.org/10.4324/9781003152651-9.
Sousa-Lara, D. 2008. Aquinas on the object of the human act: A reading in light of the texts and commentators. *Josephinum Journal of Theology* 15 (2): 243–268.

Chapter 2
Foundations of an Ethics of Excellence

Abstract This chapter lays the foundations of an ethics of excellence grounded in neo-Aristotelian moral philosophy. It clarifies the meaning of ethics as oriented to ends and human flourishing, distinguishing goods that are instrumental from goods that are ends in themselves, and situating virtue as the habituated strength that orders the former to the latter. Reframing business activity through Aristotle's distinction between praxis and poiēsis, it argues that practical wisdom (phronēsis) governs managerial action, while technē guides production, and that excellence emerges when prudence orders technique. Because human action is inherently social and political, the firm is approached as a community organized around a common purpose and common good, sustained by institutional resources yet vulnerable to corruption when instrumental goods such as money, power, or prestige are treated as ultimate ends. The chapter also considers different views of corporate responsibility—Milton Friedman's profit-centered approach, Edward Freeman's stakeholder theory, and a neo-Aristotelian perspective rooted in justice and the common good. In this vein, it develops well-being (eudaimonia) as the measure of ethical success and proposes justice—distributive, commutative, and legal/political—along with minimum and maximum standards guided by the logic of gift as criteria for responsible business practice.

Keywords Aristotelian ethics · Virtue and excellence · Practical wisdom · Action and production · Common good · Eudaimonia · Stakeholder theory · Distributive–commutative–legal justice · Responsibility · Subsidiarity · The logic of gift

Any analysis of business practices in light of Aristotle's moral philosophy demands, naturally, an initial explanation of what ethics means, and this can be done by considering the signification of ethics according to the historical context surrounding Aristotle himself and its relationship to the current signification of the word. Indeed, 'ethics' is a concept whose etymological root is found in the ancient Greek word *ethos* (ἦθος), meaning custom or behavior. Not far from this signification, the word

J. Pinto-Garay and G. Scalzo, *Justice and Corporate Excellence*, Ethical Economy 72,
https://doi.org/10.1007/978-3-032-12189-9_2

'moral', with its etymological origin in the Latin *moris*, also means custom. Ethics and morality, therefore, can be considered synonymous.

Aristotle, however, does not limit his definition of ethics to a matter of customs alone; rather, he essentially attends to the nature of human action according to which those same customs are good or bad, not *per se*, but in light of the end (*telos*) that people seek to achieve (*i.e.*, the object of their activity in what is usually done). According to this, an Aristotelian approach to business ethics could never accept, for instance, the fact that economic activities are good for the simple reason that they constitute local customs. On the contrary, business practices that constitute local customs are ethically good or bad because of the end that those who have certain customs seek to achieve with their activities. The key element of interpretation here is what makes an end of individual action a real human good and, only after such consideration, how customs support or constrain the willingness to achieve such goods.

However, the good or bad ends of personal actions are not simply a matter of philosophical or theoretical discussion in terms of simply reflecting on the morality of action. More importantly, the morality of personal actions is essentially a matter of personal growth or thriving. According to Finnis (2011) such morality is constituted by a set of basic practical principles which indicate the basic forms of personal flourishing as goods to be ethically pursued and realized. IN this sense, we understand that a person who acts for good reasons becomes a better person. *Mutatis mutandis*, ethics is a matter of human capital and the possibility of increasing personal integral development (Sison 2003). In Aristotelian terms, however, more than capital or rent, this improvement of people is properly defined as excellence and virtue. The latter, in fact, comes from the Latin *virtus*, whose root, *vis*, means 'strength,' 'power,' or 'man' (because males were seen as normally stronger in physical terms), and virtues reinforce people's character by providing a stronger capacity to act in a good or excellent manner. Thus, virtue refers primarily to habit, that is, to an inclination toward the good that a virtuous person possesses and that indicates a form of personal excellence. On the other hand, the Greek root of excellence is *areté*, which means ´that which is best´, a person's best possible performance, or 'that which is optimal in a person' for their life. In short, ethics is much more a matter of becoming a better person such that enacting the good becomes increasingly more natural.

Now, assessing excellence and virtue in any personal activity is based on a twofold categorization according to which all ends as goods can be, on the one hand, instrumental or, on the other, goods in themselves (MacIntyre 1981). Indeed, some personal ends, even though can be classified as proper goods, can be considered of secondary importance because of their instrumental value. Others, on the contrary, can be valued as ends in themselves, namely, ends that are good on their own merit and not vis-à-vis something else. Accordingly, in Aristotle terms, the purpose of an ethical way of acting is found in having the custom or habit of putting instrumental goods as a means for achieving goods in themselves. In this sense, excellence is possible only if instrumental goods such as money, power or fame are used as the conditions needed to achieve goods in themselves, such as knowledge and friendship, among others. In a different manner, if the relationship between goods and means is inverted (i.e., mistakenly identifying instrumental goods as goods in themselves), we fall into a type of activity that MacIntyre (1981) defines as corrupted. Indeed, the latter refers to activities that seek instrumental goods (such as wealth, fame, or power) as if they were the ultimate goal for action. This, in fact, can be seen in professionals who run their companies in a way that seeks efficiency and profitability as the main goals of business, replacing truly excellent human goods with other purely material ones.

2.1 Practical Wisdom

As mentioned, elaborating an Aristotelian theory of business ethics and business excellence requires a foundational understanding of the nature of human action that takes place in organizations. To do so, it is essential to consider Aristotle's understanding of human activity (*energeia*) in its threefold dimensionality, namely, contemplation (*theoria*), action (*praxis*) and production (*poiesis*) (Aristotle, *NE*, 1140a; *Pol.*, 1254a); action and production, as we will show, are most relevantly applied to business practices. Let us concentrate, therefore, on these two. Personal activity in business practices can be considered in two dimensions: *praxis*, which was translated into Latin as *actio* (*agere, actum*) and into English as action or practice; and *poiesis*, which was translated into Latin as *factio* (*facere, factum*) and into English as production. More precisely, production refers to an activity whose end is the attainment of a material good distinct from the person who produces it (an external good). For instance, shoes are a material good that is distinct from the shoemaker, and the key aspect of production is not the shoemaker´s effort, but rather the result of his/her work, that is, the shoes. Accordingly, the purpose of any production is the creation of something that has instrumental value. Practice, on the other hand, refers to an activity whose end is the good existing in the activity itself or, in other words, an internal good. For instance, studying is a practice since the knowledge gained not only remains in the student but is also proper knowledge as the student is reading and understanding the content of what he or she is studying. Naturally, both production and practice have moral value, however, the latter, *praxis*, takes precedence since the result in the acting person is more profound in terms of the achievement of better standards of excellence. Moreover, both forms of activity—whether more practical or more productive—provide an interpretive key for ethical standards in business activity; an activity mostly defined as the conjunction of productive activities (*poiesis*) and, on the other hand, managerial or strategic activities (*praxis*). In short, doing good work in a company is, for instance, both a practical and productive activity, but *praxis* has more ethical value in that it provides more chances to achieve increasing standards of excellence, particularly because of the kind of practical rationality that any practice demands.

In this way, when reframing business activities in terms of practice and production, it is essential to keep in mind the role of practical thinking and the way it relates to both forms of activity. Indeed, acting in an excellent way is a matter of rationality and each kind of activity requires specific applications of practical rationality.

Indeed, practice is specifically governed by a moral practical reason (*phronesis*)—moral intelligence we might say—and production, on the other hand, by a technical practical reason (*techné*). Now, the latter is a lesser form of practical rationality when productive tasks allow a person to follow very simple productive instructions without being particularly intelligent or deliberate. In a different manner, the practical knowledge that action requires has more complexity, particularly when deliberating and organizing means for an end in changing circumstances of action (like developing and applying a strategy). In other words, the practical knowledge proper to *praxis*—*i.e.*, strategic thinking—is a superior kind of practical intelligence often defined as practical wisdom.

Indeed, simply put, the practical reason applied to any form of practice—like managers having the needed practical wisdom to think strategically—affords the ability to (1) compare contingent goods subject to our power and open to multiple possibilities -in short, means- in order to seek the best good (practical counsel) among them; (2) be aware of our deliberative capacity, to evaluate our actions, to rectify them when they do not perfect reason, and to differentiate between what is urgent and what is important (*eubulia*); (3) judge in practical terms, that is, concerning what is probable—that which is particular, contingent, possible, and plausible, arranging the right order of things with more or less levelheadedness and experience; (4) choose—to decide among possible actions that which is most fitting; and (5) command our actions in pursuit of an end. In light of this, it becomes quite evident that any potential excellence that can be achieved in practice is substantially superior to that found in productive matters. Now, the ability to act well in practice—as we have mentioned for business strategists—provides experience and good decision-making habits, materializing excellence in what is defined as practical wisdom or prudence.

Accordingly, prudence is not the same as technical skill. There is no doubt that technique occupies an important place in managerial skills, but management, which is essentially a form of practice, involves more than an assortment of techniques. In fact, CEOs and those in leading positions usually do not have the most advanced technical knowledge within the firm, but rather a practical knowledge that, in Aristotle's definition, is economic (*οἰκονομία*) not chrematistic (*χρῆμᾰ*). Indeed, chrematistics is a technique that allows businesses to make more money. Working at a money desk, for instance, can be conceived as a chrematistic activity when applying technical knowledge that facilitates increased profitability. On the other hand, economics—more accurately translated nowadays as agency or management—is not a technique, but rather a practice that requires and relies on chrematistics or technical knowledge. Management, for instance, unfolds when choosing means amid changing conditions, with scarce resources, and more or less predictable risks. Moreover, as a practice, it develops according to human good, and not based on enriched instrumental conditions, resources, or material means, as occurs with the application of technical knowledge.

In this vein, *poiesis* is man's exercise of conceptual or technically regulated efficiency. With it, a person is capable of actively imprinting an aesthetic or useful perfection onto reality, thus bringing about what are called products of art (*techné*) as human history unfolds. Such intervention rationally transforms natural reality into man's cultural environment. In this way, humanity succeeds in constructing a human material world, that is, the material aspects of culture that we call civilization (Canals 1987). However, and despite all the technical manifestations found in society and our constant technological advancement, the technical sphere is not an active system. Instead, it is an *executive* one. When we refer to complex technical processes (like artificial intelligence for instance) we must think of them as executing rather than practical (Guardini 2000) since they do not have the deliberative capacity proper to human beings. In fact, the moral problem found in conceiving of modern companies only in terms of technical-productive-efficient systems is the risk it carries of replacing human practice and autonomy.

In this sense, business cannot be reduced to mere manufacturing because, as soon as it takes on technical aspects, it must assume the entire reality that goes along with the processes of production. These technical-productive processes as a whole surpass the mere technical aspects that go into them because what we can call technique unfolds together with a distinct, peculiar autonomy that is particular to personal action.

When human activity combines *praxis* and *poiesis* it becomes what Aristotle calls *érgon* (ἔργον). This activity is good when prudence guides technique, and when this process allows for the acquisition of skills and the virtue of prudence. This ordering between practice and production (prudence and technique) engenders a natural order, a reconfiguration of the person, a second, better nature, which we call character. In this way, a person finds her way to excellence (*areté*/ἀρετή), the highest qualitative criterion of personal activity. Thus, in matters of excellence, Aristotle does not oppose thinking to activity; he instead recognizes that all human activity—being quite complex in ethical terms—features unity between conception and execution and, more specifically, the integration of moral intelligence and technique (Murphy 1993).

On the other hand, personal activities, particularly in their practical dimension, cannot be addressed in individual terms alone. In fact, human action must be understood in light of its social reality. As Vigo (2007) explains, in Aristotelian terms, the possibility of personal fulfillment through practical action cannot be understood apart from life in community (*polis*). The social dimension of human life to which Aristotle refers as political—political praxis—is not a reflection on the social circumstances in which man generally lives, but rather a call to live oriented toward community for personal activity to be truly perfective (Aristotle, *Pol.*, 1253a). This is what we understand when Thomas Aquinas—following Aristotle's concept of *zoon politikon*—says that "*homo est naturaliter politicus, id est, sociales*" (Aquinas, *S.Th.*, I, q.96. a.4). A person is social, rather than simply living in society, and this means that any personal excellence is possible only because each one of us participates in different communities, such as a given workplace.

However, the political dimension of personal action is not only a matter of participating in a community. On the contrary, *praxis* as political action is particularly revealed in governing activities. Politics, in this sense, refers to the activities of those who lead a community, especially in terms of making possible man's practical orientation toward his ultimate end as a being that naturally lives in society. Indeed,

Aristotle himself admitted this when he affirmed that it is necessary to establish some higher end towards which all the other lesser goods are ordered when organizing the ends or goods of the different activities carried out in a community. He went on to affirm that this end must be established and recognized by a principal and eminent governing activity.

In this vein, political action is based on the consideration of the main purpose (*telos*) according to which all ambits of a community must be oriented in ambits such as technology, economics, or rhetoric. For an organization (be it a company or a city), politics is not just one more activity, but rather the first and most important activity because it illuminates the organization's ultimate end, its purpose, directing the good of the parts toward the good of the whole.

Thus, political praxis gives meaning to the action of those who make up a community. In other words, an understanding of what *we* are doing together is required to understand our own personal actions and to ensure that it is indeed rational to do in social and community contexts. In other words, politics gives fuller meaning to personal activity (Cruz Prados 1999).

However, for an organization to exist, it is not enough for a group of people to just get together. Nor is it even enough for all of them to have a common purpose. Instead, they must organize themselves and coordinate their activities, ordering their joint action toward the achievement of results that all of them consider to be in their best interests, although for different reasons. This is, in fact, the constitution of a community under the logic of a common good. According to Finnis (2011), the common good is not an empty concept, but rather a practical and tangible organizational reality that features (1) the attainment of material conditions for the development of a joint activity, (2) the coordination of actions so that the result of the joint activity is achieved, (3) and the human development of each of the members of the community through their individual action. According to this, the firm is a community of persons (Solomon 1992) acting in a social and cooperative fashion (Kennedy 2006) aimed towards providing goods and services in an efficient and profitable manner (Melé 2012) in order to achieve sustainability over time.

In short, leading as a political praxis is firstly a matter of communicating a common purpose, as well as the organization of a common good. The common end or purpose comes before the organizational process because the former is the primary cause of action and gives meaning to joint action. This end, integrated into the organizational process, is political activity *par excellence* for a good manager. However, purpose must be materialized in a common good, that is, making possible an organization according to the purpose that provides meaning.

2.2 Well-being

In the context of revisiting Aristotle's ethics to understand business activities, it is important to bear in mind that acting in a good manner—being excellent and virtuous—is justified because of happiness (*eudaimonia*). In other words, acting in a good manner is valuable any time it makes us happy. Indeed, for Aristotle, excellence and virtue are not only a force or disposition of character, but rather a way of living in which we keep thriving (Aristotle, *NE*, 1097b-1098a). In other words, happiness—that is, good living or flourishing—is a condition of personal fulfilment (Finnis 2011) and, as such, represents the ultimate standard of virtue.

Now, if the purpose of ethics is to acquire virtues and being excellent in order to be happy, this naturally applies to all dimensions of human life, like work, business, and markets. In other words, the purpose of business is happiness. However, this statement, while not incorrect, requires clarification, as the scope of happiness that business can provide is inherently partial.

To clarify this, we must firstly keep in mind that, in Aristotelian terms, happiness corresponds to the fact that human beings possess both a material existence (*physis*) and a subjective dimension in which the purpose of action (*telos*) is associated with internal goods (i.e., good in themselves as mentioned before). In this sense, happiness is a balanced conjunction between material conditions and resources (for instance, health, money, etc.) and internal goods (for instance, knowledge, friendship, etc.). Accordingly, living in such a balanced conjunction means having what we can signify as wellbeing or good living. In this vein, material conditions and resources—although not the most important aspect of good living (*i.e.*, they are not sufficient conditions for happiness—are necessary condition for having a good living and prosperity. Indeed, happiness, in its Anglo-Saxon root 'hap', means luck, favored by fortune, advantaged by circumstance, prosperity, a circumstance in which events turn out well, being fortunate. Its Latin synonymous, *felicitas*, meant, according to Cicero, pleasure, joy, good luck, a good destiny. For Pliny, it meant fecundity, fertility of the earth and plants. It comes from the adjective *felix*, which in its origin means fertile, productive, fruitful, prosperous, fortunate, rich, opulent, skillful, or propitious.

In this way, procuring adequate material conditions is not of less ethical importance, as Aristotle shows in his understanding of economics (Aristotle, *NE*, 1098b; *Pol.*, 1323a) that, nowadays, we could define as management or administration. According to this, the happiness that firms provide is much more associated with the material conditions needed for employees and commercial stakeholders to work and deal in a way that constitutes thriving. In other words, both ambits of human living—the practices of working and dealing—need a stable ordering of instrumental resources, otherwise, practices are impossible.

Firms, indeed, require an underlying layer of instrumental resources (money, power, and prestige), allowing them to work and deliver products/services to the market so that workers and customers reach higher standards of development and well-being. By way of example, to learn, one must have certain material resources, such as books. This relationship between the internal good (education) and the possession of specific external resources (the physical book) is what we call wellbeing. Now, to preserve this well-being over time (for bookstores and libraries to remain open and accessible), both the business and its customers need to be economically sustainable. In the case of a business, sustainability is achieved when it has institutional stability.

Indeed, such stable provision of material resources and conditions characterizes the institutional dimension of every community, particularly the firm. In its institutional dimension, firms are characterized by allocating the material resources (money, power, and status) needed for work and dealings to take place and, more important, to be sustained over time (MacIntyre 1981). However, the institutional dimension of the firm not only represents a condition for wellbeing, but also for corruption. Firms, indeed, become corrupted as they seek instrumental goods (money, power, and prestige) as if such goods were good in themselves (MacIntyre 2016). For instance, a firm that has promoted a culture of work in which the main purpose is nothing else but efficiency and profits, can be correctly classified as corrupt as it has neglected goods in themselves such as the personal development of workers in terms of experience, knowledge, capabilities, etc. (Pinto et al. 2021).

2.3 Business Under the Lens of Justice

Business ethics, like many areas within management studies, can be classified in what Finnis, grounded on Aquinas, defines as social theory (1998), that is, understanding, assessing, and judging both the purpose of associating and the way in which such association contributes to the well-being of the associates and the association´s common good.

In this regard, business ethics theory seeks to provide a systematic understanding of workplaces, market practices, and their intrinsic relationship with society and civil society. In other words, business ethics involves examining the relationship between the firm and its third parties—or stakeholders—within specific normative frameworks. In other words, a central aim of business ethics theory is to develop a conceptual framework through which business–society practices can be objectively evaluated. Within this context, several theoretical approaches have emerged, among which two are particularly influential alongside the neo-Aristotelian perspective: Milton Friedman's trickle-down economics, derived from neoclassical economic theory (specifically, the Chicago School), and Edward Freeman's stakeholder theory, which he describes as rooted in Kantian capitalism.

2.3.1 Stakeholders from Economic to Management Theory

On September 13, 1970, The New York Times Magazine published Milton Friedman's now-famous article articulating his position on corporate social responsibility. The article's title, The Social Responsibility of Business is to Increase its Profits, is indicative of its core message. It is fair to say that this piece—along with Friedman's broader perspective on the relationship between business and society—has become one of the most frequently discussed works in business ethics courses, largely due to its provocative and controversial stance.

Friedman argues that only individuals—not corporations—can bear moral responsibilities. As artificial entities, corporations possess only artificial responsibilities. Accordingly, to claim that firms themselves have social responsibilities is, in Friedman's view, to endorse a form of collectivism that improperly extends the political domain into all areas of human activity. If corporations are to be said to have any responsibility at all, it is, he contends, solely to use their resources and engage in activities aimed at increasing returns for their shareholders, provided they operate within the 'rules of the game'—that is, through open and free competition, without deception or fraud. Thus, the firm's sole obligation is to maximize profits for its stockholders, while complying with both legal requirements and the ethical norms embedded in society (Friedman 2007).

From this perspective, Friedman recognizes only two actors: shareholders and society. However, society plays a secondary role, benefiting only indirectly as shareholders accrue wealth and, consequently, expand their capacity to consume and invest. This logic underlies the so-called 'trickle-down effect,' which Friedman views as effective when businesses are focused exclusively on profit maximization. Accordingly, he maintains that companies should not concern themselves with issues of social equity but should instead pursue their primary function—generating profit—as a means of indirectly producing public benefit. Notably, while Friedman's account does not frame the firm's duties in terms of responsibility or justice, its minimal social consideration rests on compliance with legal obligations and prevailing social norms.

In summary, this view of corporate civic responsibility treats principles of social justice—if acknowledged at all—as separate from corporate objectives. The corporate purpose, namely profit maximization, is not necessarily aligned with ethical considerations. And when ethics do enter the equation, they tend to serve an instrumental function, contributing primarily to the firm's reputation.

In contrast to Milton Friedman's shareholder theory, stakeholder theory introduced a significant shift—a kind of new paradigm—by redefining the potential responsibilities that firms may bear toward society.

Stakeholders, by definition, are any individuals or groups that can affect or be affected by a firm's activities—either by supporting or hindering the achievement of its objectives (Freeman 1984). Although the term stakeholder had appeared in business literature as early as the 1960s (Goodpaster 1991), it gained significant prominence in management theory with the publication of Freeman's seminal work, *Strategic Management: A Stakeholder Approach* (1984). In this context, Freeman is widely recognized not only as the principal architect of stakeholder theory but also as the originator of what is now considered its classical formulation.

According to Freeman et al. (2023), stakeholder theory offers a markedly different perspective from Friedman's view on the relationship between business and society, particularly in its reconsideration of what constitutes business success. From the standpoint of stakeholder theory, success involves more than merely maximizing profits for shareholders. While profit remains a fundamental component, true success also requires delivering high-quality products and services that customers value, maintaining strong relationships with suppliers, inspiring employees, and cultivating supportive communities in which businesses can flourish.

In this regard, Freeman argues that stakeholder theory addresses the very foundation of capitalism by recognizing the broader set of third parties essential to a firm's long-term viability. It defines the responsibility of individuals in leadership positions as the duty not to undermine the foundations of capitalism, but rather to contribute to its sustainability by creating value through strategic business practices. Accordingly, stakeholder theory conceptualizes stakeholders through a dual framework: one that is both strategic-instrumental and ethical-normative.

In terms of strategy, stakeholder theory asserts that groups or individuals are not inherently stakeholders; rather, one's status as a stakeholder depends on their relevance to and functional role within the organization (Phillips et al. 2003). Social actors acquire the status of stakeholders only when they are strategically significant to the firm's goals, meaning they play an instrumental role in the organization's development. Phillips et al. (2003) further explain that managers are agents for the corporation, which is not simply another way of saying that they are agents for shareholders. Stakeholder theory, therefore, does not advocate serving two masters—and avoids the 'too-many-masters' problem that many attribute to it (e.g., Jensen 2002). Instead, managers serve the interests of one master: the organization (Phillips et al. 2003), with stakeholders playing an instrumental role in business strategy. In Freeman's words, managers bear a duty of care to the firm and are obligated to apply reasonable judgment to manage the firm's affairs in a way that attends to the interests of its stakeholders while benefiting the firm (Freeman 1994).

On the ethical or normative side, standard stakeholder theory holds that businesses must operate in ways that enhance—or at the very least, do not harm—the well-being of affected third parties. Managers are thus expected to prioritize the interests and welfare of stakeholders above the sole pursuit of profit or shareholder value maximization. These obligations arise from the recognition that stakeholders are human beings and citizens (Phillips et al. 2003), and as such, must not be treated merely as instruments of the corporation (Escudero 2010).

This ethical stance forms the basis of what Evan and Freeman (1988) refer to as Kantian capitalism—a philosophical foundation for developing stakeholder theory as a normative corporate social theory. It seeks to integrate the creation of wealth with the protection and promotion of stakeholder interests and well-being, especially those who can support or hinder the achievement of organizational goals (Phillips et al. 2003). In adapting Kant's ideas to contemporary business contexts, Freeman's stakeholder theory merges the Kantian principles of individual freedom and voluntary action (Bowie 1998) with a modern understanding of property rights (Donaldson and Preston 1995).

Purnell and Freeman (2012) explain that the normative nature of stakeholder theory is not only an abstract moral philosophy but also deeply practical. The explicit normative goals of managerial decision making—such as social value or well-being—are meaningful because they have been effectively applied in practice. This prevents stakeholder theory from becoming an empty formalism with no impact on the real world of business (Phillips 2003). In practice, stakeholder theory must be ingrained in the corporate culture. The normative aspect of stakeholder theory is therefore characterized as a set of ideas, values, and beliefs expressed by managers when talking about doing business across times and places. Indeed, Purnell and Freeman (2012) suggest that the concept of 'management talking' refers to the practice of telling stories in a threefold manner: accounting for the past of business normatively, assessing the present ethically, and projecting the future development of business to secure value that is approvable from a fitting moral perspective. Freeman and Harris (2009) argue that such narratives can be analyzed systematically to understand which principles are upheld in regular corporate practices and how these principles can be assessed as reasonable or unreasonable. In other words, theory and narratives must be integrated when corporate practices themselves play a role in validating the theory (Purnell and Freeman 2012). A stakeholder theory that cannot be integrated within a justified corporate narrative is, ipso facto, disqualified.

However, as Miles (2017) and others have noted, while proponents of stakeholder theory have offered philosophical foundations for defining stakeholders and articulating the relationship between corporations and society, these efforts have not produced a unified or systematic theoretical framework. Instead, various conceptualizations of what constitutes a stakeholder have led to multiple—and often conflicting—interpretations.

In this regard, Freeman himself acknowledges that stakeholder theory, as originally formulated, leaves several critical questions unanswered—particularly concerning the establishment of a coherent framework for prioritizing stakeholder interests and moral claims. As a result, certain business constituencies remain ambiguously defined, lacking clear stakeholder identities or distinctions (Dunham et al. 2006; Crane and Ruebottom 2011; Wijnberg 2000). This lack of conceptual precision also affects managerial decision-making and limits the theory's capacity for empirical validation across differentiated stakeholder groups (Miles 2017).

2.3.2 *Neo-Aristotelian Stakeholder Theory*

From an Aristotelian perspective, a conceptual framework for stakeholder theory can be constructed on the basis of the criteria of justice. In other words, the complex network of relationships that constitute business practices can be normatively analyzed through the lens of the Aristotelian principle of justice—particularly by considering its various parts and the gradations inherent to this virtue.

Indeed, in its definition as a virtue, justice is the good habit of 'giving to each their due' or 'giving to each their own.' Being a general criterion, it applies to all forms of relationships in human societies. But Aristotle's Book V of his *Nicomachean Ethics* shows that justice can be understood as distinguished in three parts, namely, distributive, commutative, and legal justice.

In this vein, distributive justice (Aristotle, *NE*, 1131a) concerns only those relationships between an individual and the community in which the same individual has been participating in an active manner. This is the relationship between the whole and the part, and vice versa. For this reason, it is also named distributive-contributive justice, and the criterion of justice is, in this case, merit (Aristotle, *NE* 1131a; 1131b;

Aquinas, *S.Th.*, II-II, q.61). Accordingly, distributive-contributive justice gives to each member of the firm what is due, that is, in a proportional manner considering the contribution of each single member to the common good of the community. (Aquinas, *S.Th.,* II-II, q.61, a.1-a.2).

Furthermore, justice also applies in particular to those relationships established between the firm and its commercial constituencies. According to this, the criteria for considering a commercial activity as just or fair is not simply associated with voluntary engagement, but also with proportionality in the benefits that both parties freely attain through the transaction (Finnis 2011). Freewill and proportionality are, therefore, essential features of commercial or commutative justice, that is, the justice that takes place through the interchanges and trade-offs (Aquinas, *S.Th.*, II-II, q.61, a.1); a type of justice can be seen in trading, usufruct, money lending, deposits, or rent (Aristotle, *NE*, 1131a), that is, in all actions that manifest the exchange (Aquinas, *S.Th.*, II-II, q.61, a.2). In this kind of particular justice, the principle of fairness is not merit, but mutual benefit.

On the other hand, firms establish a regular relationship through their members with civil society. This kind of relation is naturally not commercial, and it demands another criterion of justice in order to conceive what is due in business-civic matters. This justice is called legal or political and establishes the criteria of fairness applied to the relationship between citizens and private corporations with the State (Finnis 2011). Legal justice considers individual and all forms of institution in light of society as a whole, therefore, it pertains to political activity (from the *polis*), meaning everything that comes to fruition for the common good of society (which, in turn, must include commercial and productive relations).

In this context, society defines firms as corporations—organizations with legal recognition, along with duties and responsibilities toward civic society, particularly in terms of legal compliance. According to Finnis (2011), legal justice directs the actions of society toward the common good through governance; in other words, the exercise of political justice grants authorities the legitimate right to compel. This form of justice establishes priorities among the goods that human dignity and the common good demand. It determines the need to safeguard essential goods and values that are fundamental to culture and society, either by prioritizing them or by restricting secondary or relatively minor goods. For example, it is a matter of justice to recognize that maternity is a more important social good than an increase in sales, and therefore, it is just to compel maternity leave as a legal obligation for companies. Similarly, privacy and intimacy are considered more important than the right to information.

However, legal-political justice does not reclaim a total predominance of society and the State over individuals and private institutions. Political action must be in harmony with individual, family, economic, productive, labor, educational, and cultural initiatives, among others. Indeed, legal-political justice has a social-architectural role in that it often regulates spheres that are not inherently social or political without impacting the natural processes therein; however, related regulations do indirectly affect those same spheres. For example, legal justice, through its authority, can determine certain economic policies, but it must do so in a way that does not interfere with the intervention of natural economic processes, such as prices. Therefore, authorities do not determine inflation but rather increase or restrict liquidity in the markets. Such an action is thus indirectly economic. In this sense, political justice, for example, cannot oblige companies to hire, but it can stimulate them to do so.

In short, when examining the nature of a firm's relationships with its constituencies in light of an Aristotelian theory of justice, we must conclude that each firm has three kinds of stakeholders—organizational (company), commercial (market), and corporate (society)—to whom three principles of justice must be applied correspondingly (Scalzo et al. 2021), as shown below (*see* Fig. 2.1):

Furthermore, justice as a conceptual framework for assessing business practices in normative terms requires not only a distinction between its parts, but also a logic of graduality. Accordingly, we must bear in mind that justice, as a criterion for business quality, cannot be conceived as a fixed standard. On the contrary, in matters of justice there is always a possibility to improve, even in situations in which there are acceptable standards of justice. This means that there is always a chance to improve the reciprocal contribution between the firm and its stakeholders in order to increase the wellbeing of workers, consumers, and civic society. For this reason, it is also important to conceive the relationships between business and stakeholders under the criteria of minimum and maximum justice, this is, as a range of graduality.

For many firms, for instance, minimum justice simply means complying with law or local customs in terms of remuneration or selling products or services. Maximum justice, in a different manner, means trying, as much as possible, to go beyond the

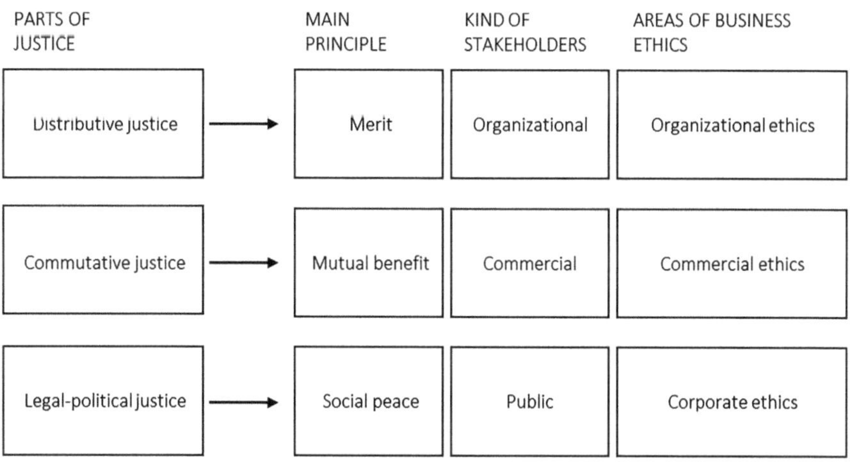

Fig. 2.1 Flow chart illustrates parts of justice and their relation to business ethics. In particular, the first column shows the three parts of justice—i.e., distributive, commutative, and legal-political—and, according to the flow shown in arrows, their corresponding principle of justice according to which the firms establish specific forms of relationships with organizational, commercial, and corporate stakeholders according to the principles of merit, mutual benefit, and social peace respectively

minimum requirements set out by law, like improving working conditions and product quality, among other initiatives. In fact, acting based on a criterion of maximums is what we call responsibility in terms of doing the best in order to reach the good of others or the common good in accordance with the logic of gift, the initiative to invest in the relationship that firms already have with its members and stakeholders (Scalzo et al. 2023).

In this vein, gift is a provision of goods and services without obligation, guarantee, or certainty of return made with the intention of creating, maintaining, or regenerating a social relationship. In other words, this is giving without return, not in the sense that rewards are discarded, but in the sense that rewards are not the principal motivation. In fact, giving should cause gratitude in those who have benefited, and this can in fact build or strengthen a relationship, and, therefore, an expression of fraternity and benevolence. Therefore, the practical reason contained in the logic of gift makes it possible to overcome the logic of individualism and selfishness, which do not really contribute to the development of society, companies, and markets (Scalzo 2017).

Indeed, as a social virtue, maximum justice involves a constant commitment to giving each person what they are due. It manifests in degrees, ranging from simply avoiding harm and not actively causing injury to others, to actively seeking their well-being. Inspired by charity, justice helps foster stronger and more lasting relationships. As it develops, I not only give others what is rightfully theirs, but also share what is 'mine.' In this sense, justice of maximum represents the principle of responsibility in business, market, and social matters (Scalzo et al. 2018).

References

Aquinas, T. (1274/1947). *Summa theologica (fathers of the English Dominican Province, Trans.)*. New York: Benziger Brothers.

Aristotle. 1995a. Nicomachean ethics. In *The complete works of Aristotle*, ed. J. Barnes, vol. 2. Princeton: Princeton University Press.

Aristotle. 1995b. Politics. In *The complete works of Aristotle*, ed. J. Barnes, vol. 2. Princeton: Princeton University Press.

Bowie, N. E. 1998. A Kantian theory of capitalism. *Business Ethics Quarterly* 8 (1): 37–60.

Canals, F. 1987. *Sobre la esencia del conocimiento*. Barcelona: PPU.

Crane, A., and T. Ruebottom. 2011. Stakeholder theory and social identity: Rethinking stakeholder identification. *Journal of Business Ethics* 102 (1): 77–87.

Cruz Prados, Alfredo. 1999. *Ethos y Polis. Bases para la reconstrucción de la filosofía política*. Pamplona: EUNSA.

Donaldson, T., and L. E. Preston. 1995. The stakeholder theory of the corporation: Concepts, evidence, and implications. *Academy of Management Review* 20 (1): 65–91.

Dunham, L., R. E. Freeman, and J. Liedtka. 2006. Enhancing stakeholder practice: A particularized exploration of community. *Business Ethics Quarterly* 16 (1): 23–42.

Escudero, G. 2010. *Bien común y stakeholders. La propuesta de Edward Freeman*. Pamplona: Ediciones Universidad de Navarra.

Evan, W. M., and R. E. Freeman. 1988. A stakeholder theory of the modern corporation: Kantian capitalism. In *Ethical theory and business*, ed. T. Beauchamp and N. Bowie, 97–106. New Delhi: Prentice Hall.

Finnis, J. 1998. *Aquinas: Moral, political, and legal theory*. Oxford: Oxford University Press.
Finnis, J. 2011. *Natural law and natural rights*. 2nd ed. Oxford: Oxford University Press.
Freeman, R. E. 1984. *Stakeholder management: Framework and philosophy*. London: Pitman.
Freeman, R. E. 1994. The politics of stakeholder theory: Some future directions. *Business Ethics Quarterly* 4 (4): 409–421.
Freeman, R. E., and J. D. Harris. 2009. The new narrative of business. *Business Ethics Quarterly* 19 (2): 233–239.
Freeman, R. E., J. S. Harrison, A. C. Wicks, B. L. Parmar, and S. de Colle. 2023. The problems that stakeholder theory tries to solve. In *R. Edward Freeman's selected works on stakeholder theory and business ethics*, 3–27. Cham: Springer International Publishing.
Friedman, M. 2007. The social responsibility of business is to increase its profits. In *Corporate ethics and corporate governance*, ed. W. C. Zimmerli, M. Holzinger, and K. Richter, 173–178. Cham: Springer.
Goodpaster, K. E. 1991. Business ethics and stakeholder analysis. *Business Ethics Quarterly* 1 (1): 53–73.
Guardini, Romano. 2000. *Ética, Lecciones en la Universidad de Munich*. Munich: BAC.
Jensen, M. C. 2002. Value maximization, stakeholder theory, and the corporate objective function. *Business Ethics Quarterly* 12 (2): 235–256.
Kennedy, R. 2006. The goods of work, distributism, and Catholic social teaching. *Journal of Catholic Social Thought* 3 (2): 277–300.
MacIntyre, A. 1981. *After virtue*. Notre Dame: University of Notre Dame Press.
MacIntyre, A. 2016. *Ethics in the conflicts of modernity: An essay on desire, practical reasoning, and narrative*. Cambridge: Cambridge University Press.
Melé, D. 2012. The firm as a "community of persons": A pillar of humanistic business ethos. *Journal of Business Ethics* 106 (1): 89–101.
Miles, S. 2017. Stakeholder theory classification: A theoretical and empirical evaluation of definitions. *Journal of Business Ethics* 142 (3): 437–459.
Murphy, J. B. 1993. *The moral economy of labor: Aristotelian themes in economic theory*. Yale: Yale University Press.
Phillips, R. 2003. *Stakeholder theory and organizational ethics*. Oakland: Berrett-Koehler.
Phillips, R., R. E. Freeman, and A. C. Wicks. 2003. What stakeholder theory is not. *Business Ethics Quarterly* 13 (4): 479–502.
Pinto, J., I. Ferrero, and G. Scalzo. 2021. Virtuous work and organizational culture: How Aristotelian practical wisdom can humanize business. In *Human centered organizational culture*, ed. M. T. Lepeley. Routlegde: Routledge's Human Centered Management Book Series.
Purnell, L., and R. E. Freeman. 2012. Stakeholder theory, fact/value dichotomy, and the normative core: How stakeholder theorists can teach business ethics. *Journal of Business Ethics* 109:109–116.
Scalzo, G. 2017. A genealogy of the gift. In *Perspectives on philosophy of management and business ethics: Including a special section on business and human rights*, 31–45. Cham: Springer International Publishing.
Scalzo, G., A. M. Almárcegui, and M. D. L. Á. Padilla. 2018. Rebuilding the temple of graces: Gift-giving as the foundation of care. *Œconomia. History, Methodology, Philosophy* (8–2): 139–155.
Scalzo, G., J. Pinto-Garay, and K. Akrivou. 2021. CSR and virtue ethics: The common good of firms, markets, and civil society. In *The Routledge companion to corporate social responsibility*, 78–87. Routledge.
Scalzo, G., K. Akrivou, and M. J. Fernandez Gonzalez. 2023. A personalist approach to business ethics: New perspectives for virtue ethics and servant leadership. *Business Ethics, the Environment & Responsibility* 32:145–158. https://doi.org/10.1111/beer.12435.
Sison, A. J. G. 2003. *The moral capital of leaders: Why virtue matters*. Edward Elgar Publishing.
Solomon, R. C. 1992. Corporate roles, personal virtues: An Aristotelean approach to business ethics. *Business Ethics Quarterly* 2 (3): 317–339.

Vigo, A. G. 2007. *Aristóteles*. Las Condes: Instituto de Estudios de la Sociedad.
Wijnberg, N. M. 2000. Normative stakeholder theory and Aristotle: The link between ethics and politics. *Journal of Business Ethics* 25 (4): 329–342.

Chapter 3
Organizational Ethics

Abstract This chapter examines the firm as a community of persons in light of Aristotelian virtue ethics, with the goal of offering a foundation for understanding the nature of work within organizations. It begins by revisiting classical theories of the ethics of work, highlighting the dignity and dimensions of good work, within a framework of rights and duties ordered by distributive justice and the principle of merit. Building on this foundation, the chapter addresses applied aspects of workplace ethics, providing ethical grounding for common challenges such as hiring, leadership, and fairness in compensation. It then turns to the responsibility of business administrators, focusing on the professional nature of management and its role in sustaining the firm as an institution. Justice serves as the guiding principle for evaluating the relationships between different constituencies, particularly through proportionality in contribution and distribution. Special attention is given to the fiduciary responsibility of managers in modern corporations where ownership and administration are separated. This responsibility, it is argued, requires an anthropological foundation stronger than classical agency theory and is best understood through the virtue of magnanimity, conceived as a form of personal excellence and ambition oriented to the common good. The chapter ultimately argues that conceiving the firm as a community governed by virtue and justice enables a more robust account of its ethical responsibilities, integrating economic performance with human development and the pursuit of excellence.

Keywords Organizational ethics · Distributive justice · Human flourishing · Personal work · Working conditions · Power and authority · Leadership · Management ethics · Fiduciary responsibility · Agency ethics

Virtue ethics describes the firm as a community conformed by workers and managers—those who provide intellectual and productive outcomes though their tasks—along with investors or shareholders who supply the financial resources for the firm to operate (Sison and Fontrodona 2012, 2013; Melé 2012) in a way in which their economic returns depend on the firm´s performance, unlike in the case of creditors. Thus, shareholders, managers, professionals, and operational employees assume

J. Pinto-Garay and G. Scalzo, *Justice and Corporate Excellence*, Ethical Economy 72,
https://doi.org/10.1007/978-3-032-12189-9_3

their respective rights and duties as members of the same community (Argandoña 1998). And such rights and duties are derived from a scheme of justice in contribution-distribution that is meant to be applied under the logic of the principle of merit. Here, Aristotelianism can be helpfully combined with (some) standard economic reasoning according to which different productive inputs (e.g., labor, capital) should be compensated according to their marginal contributions. But notice straightaway: There is conceptual room for 'merit' to pull apart from 'marginal contribution'—and Aristotelianism offers a valuable, virtue-based way to conceive of it. Perhaps human excellence, here meaning possession of the virtues, is one way to determine 'merit,' quite apart from the demand-driven considerations of marginal contribution.

Aquinas relatedly explains that "something is given to a private individual, in so far as what belongs to the whole is due to the part, and in a quantity that is proportionate to the importance of the position of that part in respect of the whole" (Aquinas, *S.Th.*, II-II, q.61, a.1-a.2). 'Position' here, as described by Aquinas, means the importance of an individual in achieving the communities' outcomes, that is, the person's performative merit. According to Finnis (2011a), the criterion of merit lies in the effective contribution of each firm member in a co-dependent, proportional manner. Therefore, according to such criterion, the distribution of benefits *cannot be equal*—otherwise it would be a form of *injustice*! After all, organizations receive different qualitative and quantitative contributions, especially in terms of work (Gomberg 2016).

Alasdair MacIntyre, in his neo-Aristotelian perspective, accordingly, holds that the common good of the firm is a matter of contribution and personal development. "The common goods of those at work together," writes MacIntyre, "are achieved in producing goods and services that contribute to the life of the community and in becoming excellent at producing them" (2016, p. 170). This is one reason why the firm's responsibility to organizational stakeholders can be defined as supporting work in such a manner that each member of the organization could improve as a human being through his or her engagement in productive activities (Sison and Fontrodona 2013). As explained by Sayer (2009), the distributive justice approach, instead of considering individuals as recipients, emphasizes the importance of self-fulfillment and wellbeing by means of each member of the firm's effective contribution to the firm's common good. Implicit in this is the idea that doing is more important than having and, indeed, what we do is what we become; we choose who we are, and we are what we choose. Our repetitive practice of choosing forges our character even as our previously forged characters influence our patterns of choice.

The distributive relationship between the firm and its organizational stakeholders is thus a matter of personal development through making specific contributions to the firm. Contributions have merit in accordance with the value of their influence on organizational outcomes. So even though personal work has the same ontological value and dignity for each employee, some kinds of work are more important than others depending on their differential impacts on firm performance. This is often, for instance, the case for those who play leading (positive) roles in corporate governance. Their work has more merit—but not dignity-based value—than that of corporate employees with smaller positive impacts (we will explain this argument further in Sect. 3.2.1).

3.1 Personal Work

Work has brought about one of humanity's great questions for as long as we have had a historical record of civilization. Like matters of divinity and the supernatural, the creation of the world, or the meaning of death, the idea of work—its importance or its meaning—is essential for reflection on human existence. In this regard, one of the oldest known considerations of the meaning of work is found in Genesis, which in fact identifies this human activity, work, with humanity itself: each person was made to work (*ut operaretur*).

Despite its radical importance, a definition of work is still challenging, particularly because of the many contemporary ways such activity has been referred to, emphasizing, therefore, some specific features of what work means nowadays. For example, work has been defined simply as 'occupation,' referring to the time dedicated to a task within a specific, delimited period. Having an occupation—being occupied—implies a focused commitment to a particular task, as seen in the case of a guard. Similarly, the term 'job' is often used synonymously with work. Its etymology, though unclear, dates back to the sixteenth century, where it initially meant a task or piece of work.

In a different manner, we usually understand work as an 'employment,' that is, work being more than an occupation as it relates to the contractual or economic aspect: the agreement between two parties to engage in a mutually beneficial economic relationship, often over an extended period or based on results. In another sense, work can refer to a trade or craft—a specific set of skills honed through experience and recognized by society through its outcomes and excellence. A trade, therefore, is defined not only by technical knowledge but also by proven exceptional performance, as exemplified by a pianist. On the other hand, work as 'labor' carries a distinct connotation, both in common usage and in philosophical terms. As noted, and following the ancient distinction, labor is considered the work of the body, rather than those who work with their hands—craftsmen or *cheirotechnes*—(Arendt 1958). The latter refers to production, as discussed earlier. In contrast, labor refers to service—it does not produce something new that endures over time, like a pair of shoes. As a service, work refers to the activity aimed at maintaining the conditions of use in an incidental manner, serving the needs of others, as with cleaning or cooking. In this context, bureaucratic jobs can also be classified as labor, especially when they involve sustaining administrative processes whose results do not persist over time.

Yet, despite the different ways in which human work is defined, it is important to conceive a common definition, particularly in ethical terms. Indeed, an ethical consideration on work outlines a transversal feature, namely, the way each form of work provides an opportunity to enhance the workers' thriving or, in Aristotelian terms, their excellence and virtue.

What does it mean to understand human work in terms of virtue and, particularly, according to an Aristotelian definition of it? In Aristotle, the concept of work can be considered an inherently human function, *ergon*, and therefore understood as a perfective action. Human action is described as perfective, and an act of work cannot be excluded *a priori*, for it would cease to be truly human or an act involving reasoning or logos. *Ergon* is a concept that describes work as a truly human function in Aristotle's philosophy, and it refers to human action in general. Such actions may or may not be perfective or characterized by excellence (*areté*), depending on whether they involve reasoning (*logos*) and whether they are ordered by reason. Human activity can be productive (*poiesis*) while being guided by reason (*logos*) in what we have explained as practical activity (*praxis*) making it possible for all human actions to be functional within the context of rationality (Gómez-Lobo 1989). Therefore, 'work' as *ergon*, in order to be excellent is much more a conjunction of *praxis* and *poiesis*, that is, being deliberative and practical in productive tasks.

In this vein, the practical dimension of work is what opens the possibility for this human activity to be an opportunity for personal thriving by means of reasoning on practical matters at work, that is, deliberating and providing meaning and significance to our daily tasks at work. Reasoning as an essential dimension of 'work' is, indeed, crucial for advancing a definition of work as a distinct type of action because of its meaning, which is different from the meaning of other activities like playing, going to school, socializing, and so on. Reasoning, indeed, provides meaning not only to our personal activities in a general sense, but also in particular, giving each activity its own significance. Playing, for instance, is not the same as working because each activity carries different purposes and, accordingly, different significations for those who perform. Accordingly, some of our activities can be properly defined as 'work' when they are provided with a specific signification from the person who is doing it. In this sense, the first way to classify a personal action as 'work' comes *ad intra*, not *ad extra*; subjectively, not objectively. To put it in another way, what characterizes work is our capacity to signify such action as 'my own work,' not necessarily complying with what a society or a culture has defined as work. In fact, this is the reason why a household chore, even though not being directly paid or made under any contract, is properly a kind of work if the person who is dedicated to such tasks considers it is his or her own work.

Therefore, work is not defined as such because of salaries, contracts, having an employer, or just because society has recognized it as work. Undoubtedly such aspects of work are quite important. However, the main reason why personal action is primarily work is because it is conceived as such by that same person who works. Therefore, the activity we call work does not depend so much on verifying the economic, productive, or objective dimension of work, rather, it depends primarily on a criterion that comes from the internal or subjective dimension, that is, from giving a sense of work—a narrative meaning—to our daily tasks. In fact, the possibility to provide intelligibility and meaning to personal actions depends on our own capacity to narrate 'what I am doing,' 'what I have done,' and 'what I'm willing to do.' This, as we have mentioned before, refers to the practical dimension of action (*praxis*), nut this time, in terms of a narrative (MacIntyre 1981)

However, providing a specific form of narrative or meaning to our working activities, even though essential, is not enough in classifying certain activities as human work. Indeed, the definition of work as a form of excellence must be articulated with two additional criteria, namely, the circumstances of work—most of them obligatory, adverse, and difficult—and what it means to perform a good job.

3.1.1 The Unavoidable Adversity

The idea of working as a condition for personal fulfillment can easily 'clash' with reality. Most people who work, regardless of their culture, country, or level of economic development, must contend with subsistence, urgency, tiredness, and uncertainty—three conditions that make work seem more like a form of condemnation than a space for personal development.

In terms of subsistence, work often carries a sense of urgency. We all must procure what we need to eat, shelter and clothe ourselves, rest, etc. The human condition, in fact, compels us to meet daily needs that can be life-threatening if left unmet. Thus, work, which is naturally associated with procuring the goods needed to satisfy basic needs, can never be detached from its obligatory nature, stemming from the necessity of seeking what is essential for living. In this sense, work inherently implies an obligation.

However, this obligatory nature extends beyond the mere satisfaction of basic needs; it also arises from the social and cultural aspects of personal life that are mostly related to what is called a human way of living. Subsistence is accompanied by demands related to lifestyles. Regarding the social demands of work, we must recognize that people live according to a cultural heritage, with moral traditions and standards of living that implicitly or explicitly define certain needs as obligatory, even if they are not necessarily vital. This arises from the appraisal of our historical condition, which we associate with being members of a family or a community. According to this, we regard certain 'minimums' related to our standard of living as essential, not only for ourselves but also for those who depend on us (such as future generations). These include the need for education, vacations, art, fashion, and social gatherings, among others. All of these 'cultural standards' contribute to the sense of non-physiological obligations associated with work, as it enables us to maintain a certain way of life.

Moreover, work is not only intrinsically associated with human obligations, but it is also subject to a condition of uncertainty. This uncertainty originates from the realm of needs—not only in terms of consumption that depletes resources not always available but also on the fact that there is almost never enough to guarantee the satisfaction of future needs, particularly of unexpected and unforeseeable events that demand extraordinary resources for a good human living. The experience of work, in this sense, is quite dominated by the logic of uncertainty—a certain insecurity that must be confronted and, if possible, resolved. For this reason, work is closely associated with savings and planning, not because work itself provides guarantees of certainty, but because it allows us to partially protect ourselves from the unforeseen.

In fact, the obligatory condition of work is present in its earliest definitions, particularly in its Latin use, *trepalium*, referring to an instrument of torture. *Trepalium* comes from *tripalis*, which means 'to put three sticks.' In this line, work has its root in the Anglo-French verb *travailler*, which originally meant to torment, then to bother, then to journey, and then to travel. Hence the reference to work as toil and trouble, whose meaning is obviously pejorative. And before that, such a categorization of work as a negative activity has its origins in Ancient Greece, in the fourth century B.C., when a sharp distinction was made between servile activities performed by slaves and the liberal arts. In this context, work referred only to the servile tasks of slaves. In fact, being freed from all servile labor was considered a privilege, and the Greeks had a special term for this freedom: *skolé* (the equivalent of the Latin *otium*), from which we derive the concepts of school and leisure, referring to activities that are thriving and non-obligatory, done with liberty, liberal, outside the ambit of work. These activities were the privilege of Greek citizens and Roman patricians, free from the condemnation of daily survival. Work (*skolé, neg-otium*), on the other hand, was the domain of animals or slaves; a sort of conviction that slaves were obligated to assume for them and for their masters, particularly the works defined labor, needed to be done in a permanent manner, like Sisyphus, the character from Greek mythology who had to eternally roll a rock up a mountain because it rolled towards the base every time he reached the top.

However, the perception of work as exclusive to slaves changed with the emergence of the Judeo-Christian tradition in the Mediterranean during the first century AD. This new understanding of work, while realistic, integrated the conditions of subsistence, obligation, and uncertainty as a condition for personal development. In fact, according to such a perspective, the conditions allowed work to become a thriving activity for all members of a community, not just the citizens and patricians. Rather than being a condemnation, this new approach to work transformed adversities into meaningful challenges, particularly in light of what we can describe as the natural meaning of work and our own personal capacities, abilities, or tendencies—what is usually called the vocational sense of work.

3.1.2 Morally Meaningful Work: From Friendship to Vocation

The instrumental condition of work—according to which almost every form of human labor is aimed at securing subsistence—is, at least, incomplete. There is no doubt that human work can rightly be described in terms of utility and instrumentality, insofar as labor enables us to provide ourselves with the material conditions for living. However, this basic condition of work cannot be its sole characteristic, because work, if it does not overcome its sole condition for subsistence, will never be a sphere of personal excellence and common good beyond mere qualities of efficiency. In contrast, for work to be conceived in moral terms, it must be intrinsically directed toward goods in themselves, something akin to the intrinsic goods of work according to which it is possible to provide meaning to mostly any kind of toil—not, however, as in *skolé*, outside working practices, but within them. Where can a sense of excellency in work be found in order to overcome its instrumental condition? Which is the good in itself that could give meaning to work beyond the mere need for survival?

Consistent with the political nature of human beings—i.e., what Aristotle defined as *politikon zoon*—work must ground its condition for excellence in a twofold condition: work is always carried out 'with others'—work is never entirely isolated—and that the only good in itself is the good to be found in 'others,' since the only good in itself is a person. Accordingly, the possibility of elevating work beyond its instrumental dimension for subsistence into a condition of excellence depends on integrating a sense of friendship as the primary reason for working. In other words, the intrinsic good that can give meaning to work is to be found in what Aristotle calls perfect friendship.

According to Aristotle (*NE*, 1155b-1156a), friendship can be distinguished into three types arranged in a kind of graduality and hierarchy: from the least valuable form, which is friendship for instrumental reasons, to friendship based on mutual pleasure or satisfaction, and finally to the most perfect form of it, which is personal friendship grounded on virtue.

> Perfect friendship is the friendship of men who are good, and alike in virtue; for these wish well alike to each other qua good, and they are good themselves. Now those who wish well to their friends for their sake are most truly friends; for they do this by reason of own nature and not incidentally; therefore their friendship lasts as long as they are good-and goodness is an enduring thing. (Aristotle, EN, 1156b)

In this vein, Finnis (2011a), following Aristotle, explains that friendship (*philia*) is not always perfect, but it can be manifest partially in friendships *secundum similitudinem*, that is, relationships of utility and/or pleasure or satisfaction that do not amount to full virtuous friendship. According to Mulgan (1999), perfect friendship in Aristotle is essentially characterized by both the intention and the effective demonstration of goodwill (*eunoia*) in a reciprocal manner, that is, mutually intended between friends. This implies that each friend seeks the good of the other for the other's own sake, rather than for personal advantage (Aristotle, *EN*, 1166). In other words, perfect friendship is marked by each person acting for the wellbeing of the other as if it were their own, valuing the other's good as the foundation and the cause of their own wellbeing, all within a relationship of reciprocity. In more practical terms, this is working based on the Golden Rule principle according to which we assume the responsibility to 'treat others as we wish to be treated' or, conversely, 'do not do to others what we do not want to be done to us' (Finnis 2011b).

Even though *eunoia* is absent in friendships motivated solely by utility or pleasure (Aristotle, *EN*, 1241a4–7; 1167a12–13), perfect friendship does not exclude utility or pleasure, but it rather can integrate them within a context of reciprocal goodwill—it opens the possibility of transforming, for instance, business relationships based merely on individual utility into friendly dealings where people take into consideration their counterparts' welfare without neglecting personal utility (Mulgan 1999).

In this sense, Finnis (2011a) explains that a definition of work in terms of friendship is not in contradiction to the fact that labor partners or contracting parties usually establish in their initial stages of relationship based in utility. As such, they operate on the basis of limited common interests, structured under more or less coordinated actions aimed at achieving some goal that might be valued for purely individual interests. According to this, the benefit of any others labor partner is a matter of indifference (unless that success impacts one's own interest), the same way that the coordination of working activities is valued only insofar as it yields an individual benefit.

As mentioned, working for reasons of utility or pleasure/satisfaction cannot be clasified as false friendship, even when they do not reach the highest form of activity according to Aristotle's concept of perfect friendship—*i.e.,* one based on reciprocal goodwill. Rather, they are better understood as partial friendship, as they can usually be open to the possibility of attaining a moral sense beyond mere utility and pleasure. Indeed, work, understood under the logic of friendship, can transcend its purely instrumental condition for subsistence by integrating the need to secure the means of living into a higher reason for action, namely, working for the benefit of others. Only by embracing this moral meaning can work truly become an opportunity for excellence and the common good without disregarding the importance of subsisting and having what is needed for a good living.

Moreover, the need to work for others as condition for being excellent is not a matter of idealism, a way of abandoning pragmatism. By contrast, Finnis (2011a) argues that if such a moral dimension of excellence in perfect friendship is excluded, work becomes unreasonable, as it forecloses the possibility of contributing to both personal and common good—that is, to the basic human goods without which our own human flourishing (*eudaimonia*, in Aristotle's terms) is impossible.

However, the sense of friendship in matters of work must be also conceived in light of the uniqueness and originality derived from the dignity of the human person (*see* further on Sect. 3.2.1). In other words, the way in which we value the reciprocal goodwill friendship must be marked by the fact that in such relationships we should act in accordance with our own capacities, abilities, tendencies, tastes, in sum, our own personal way of working in what is defined as vocation.

Etymologically, the word 'vocation' derives from the Latin *vocatio*, meaning 'a calling' or 'being called'—vocation often begins with a reflection on work in relation to religion. The etymological definition highlights the idea that we are, in essence, called by our Creator, making vocation inherently a matter of faith. However, while this transcendental perspective is correct, there is also a philosophical aspect to vocation that pertains to 'who we are' as individual persons. Philosophically, vocation encompasses a natural understanding of our unique characteristics according to which we establish relationships of friendship. Each of us can find meaning in our work based on our original history of living, abilities, tastes, interests, and opportunities. In this sense, vocation relates to the creation of our personal biography—how we shape our work in ways that resonate with our individual experiences and values, even in the face of adversity.

In this sense, vocation becomes a challenge to make our work a deeply personal matter. Finding a vocation involves reflecting on the activities we engage in at work, their intrinsic human value, and their usefulness. It is an understanding of our own capacities, abilities, and disposition, directed both toward our personal well-being and the well-being of others.

Thus, discovering a vocation enables us to identify personally with our occupation, employment, or job. It allows us to recognize that, through these activities, we express who we are, how we do things, for whom we do them, and how our work serves others (Scalzo et al. 2023). In this way, each of us is called to find our own unique vocation—a mission in life—not waiting for the meaning of life to be handed to us but actively seeking the meaning of our working life as it aligns with the broader purpose of our existence (Frankl 1985).

However, the problem with such a definition of work in terms of vocation is the fact that most of the workers in the world are not involved in activities that they might have chosen as vocational. We can suppose that most of the world's working force is indeed doing simply what they can in order to have a basic living, not what they have freely estimated as the best activity according to their vocation. Vocation, in this sense, would be nothing more than a good aspiration, but quite unrealistic. Is this correct? How can each of us find a vocation even in those occupations that we have not chosen?

The vocational sense at work requires, in our opinion, distinction between the functional dimension, on the one hand, and its dispositional features, on the other. The former can be evidenced in certain aspects associated with know-how, technical knowledge, or specific skills associated with specific functions. In other words, any vocation can be identified with certain functions, such as software programming, nursing patients, driving a police car, and so on. In fact, we might say that children identify their vocation mostly with this functional sense.

In another sense, vocation involves certain dispositional traits, characterized by a sense of authenticity in one's personal capacities and tastes, reflecting unique and individual tendencies in the most subjective aspects of work according to which we seek the benefit of others in very authentic ways. In this regard, having a vocation aimed at caring for the security of other people and the vocation for driving a police car are two distinct things. The latter is a function, while the former is a deeper calling based on our personal traits that can be put into practice in different functions.[1]

Yet, to become a source of personal development, the dispositional dimension of a vocation requires certain virtues to deal with what we have explained as the adversities of work. Indeed, the vocational sense of work sheds a positive light on the weariness, difficulty, and effort proper to all work. For Aquinas, work, like other

[1] In this vein, when any vocational reflection confuses the moral meaning of work, there is a risk of moral deterioration or corruption. This is, in fact, the case in which the meaning of work is wrongly settled in the functional dimensions of a vocation, instead of its correct moral sense. By way of example, this is the case of those who have assumed their vocation as making money, wrongly conceiving their work as meaningful as its reported success, efficiency, or rent. Therefore, the only way to assess the value of work is, for instance, a certain position in a company, a certain income level, economic achievements, or recognition. However, this is a form of corrupted meaning of work; something quite risky, particularly when the circumstances turn adverse, no success is achieved, work becomes nothing more than a complete disaster, a rupture in life. In many cases, such risks require assurance and control. This may explain why many modern workers, particularly those in leadership positions, often have an unrealistic ambition for accumulation and control. Common sense, on the contrary, tells us that there will always be adverse circumstances, drastic changes, uncertainty, and undesirable outcomes.

human activities, is better insofar as it involves the virtue of industriousness, that is, the habit of achieving arduous goods (*bonum arduum*: Aquinas, S.Th., I-II. Q.40.a.1). In other words, the virtue associated with effort and difficulty inherent to any human work. Something similar we might say about the virtues of patience and peace, that is, the virtues aimed at conserving a positive sense of work despite the effort, fatigue, and uncertainty involved. In other words, the dispositional features of a vocation need virtues to be put into practice.

However, it is still quite reasonable to consider that many workers do not have the opportunity to have a job in which their functional or dispositional features of a vocation can be displayed. In such cases, however, even when a vocational sense of work—functional or dispositional—is out of any possibility, we can still rely on the moral meaning of work; a moral meaning that, in fact, should be also present in any vocational appraisal.

In simple terms, such a moral sense of work is defined by the importance of seeking both a personal development and a contribution to other people´s good in an integrated manner, that is, conceiving the good of others as my personal good and vice versa. This, in fact, can make any activity—vocational or non-vocational—morally meaningful, no matter the circumstances.

According to Viktor Frankl (1985)—a Jewish psychiatrist deported to a concentration camp during World Warr II[2]—if we are not able to change our circumstances or the situation in which we are living, then we must face the challenge of changing ourselves. This implies a responsible search for meaning. Hence, the meaning of life, even in very difficult circumstances, is a personal responsibility that arises from the exercise of freedom: There are no circumstances, no matter how bad they may be, that exonerate us from acting freely. In this context of morally meaningful work, fatigue, survival, or uncertainty become less determining of work.

Hence, adverse circumstances can be overcome in light of appreciating one's own development and the common good. Frankl himself writes that those who survived the concentration camp always remember fellow prisoners who walked among the bunks comforting others or giving them their last piece of bread. Following Frankl's idea, we can say that adverse circumstances cannot eliminate the capacity to provide moral sense to any kind of occupation, turning it into a meaningful activity.

In summary, in terms of moral sense and vocation, modern work presents a prisoner's dilemma—though not the one typically discussed in microeconomics. This 'prisoner,' akin to the one described by Frankl, represents modern workers who often face adverse and compulsory circumstances. However, these workers always have

[2] Born in Vienna to a family of Jewish origin, Frankl was an important twentieth-century neurologist and psychiatrist. He studied medicine at the University of Vienna and specialized in neurology and psychiatry. Before World War II, he worked at the Vienna General Hospital and had a private psychiatry practice. He also headed the neurology department of the Rothschild Hospital (the only hospital in Vienna where Jewish patients were admitted). In September 1942, together with his wife and parents, he was deported to the German concentration camp Theresienstadt. He then spent time in four concentration camps, including Auschwitz, and, although he managed to survive, he lost his wife, family, and friends.

the capacity to imbue their occupations with moral meaning—i.e., a sense of perfect friendship or, in other words, working for the sake of other.

3.1.3 Dissecting the Well-Done Work

So far, we have analyzed work from a personal perspective. However, from the standpoint of work ethics, it is essential to evaluate work through the lens of an impartial spectator. This, in our opinion, involves understanding what makes a specific job qualify as good when viewed from an external perspective. This, in fact, enables us to assume a more 'business ethics' perspective while assessing, for instance, the moral quality of work in Asian factories, high tech laboratories, or that of guards in surveillance companies or musicians in a national orchestra.

According to our Aristotelian approach, any form of work, *ergon*, must comply with being good in four complementary aspects, namely, productive, practical, cooperative, and benevolent.

A first *sine qua non* condition for any work to be considered well-done is productivity. In line with what we have said before, such a productive dimension is philosophically defined as *poiesis*, that is, an activity that is aimed at creating products—manufacturing—or, in other cases, material changes as it happens with labor or service.

In terms of quality, what is important in productive activities is not who performs, but the product or result itself. Accordingly, the activity of the producer is valued in light of the product, that is, in terms of efficacy and efficiency. The latter, in fact, refers to a decreasing protagonism of the producer that does not affect the quality of the product or, in some cases, can, in fact, improve it in a producer-product inversely proportional relationship defined as being more productive. The important aspect of production is, therefore, the product as, for instance, the case of a shoemaker who can be qualified as a good one only if shoes are good products, not because of his or her intentions, experience, knowledge, etc. Likewise, a shoemaker can be called an efficient one only if the processes of production, being the same, obtain better shoes or, in a different manner, if the process of production, while demanding less time and resources, obtains the same kind of shoes or better ones.

In this sense, the practical reasoning needed to produce, namely, technical knowledge and skills, are also valued in light of the product. This can be argued just by considering the fact that some techniques and skills, no matter what their complexity and difficulty to obtain, are obsolete, that is, they do not have any useful purpose as the products associated with them are worthless.

That said, the product itself is worthy in light of two criteria: instrumentality and aesthetics. The former is quite simple to understand: the quality of a product is defined by its capacity to obtain something else (i.e., instrumentality) in a way in which its usability is simple. In other words, a product is a good one as it represents an inverse proportional relationship of instrumentality and usability.

Additionally, products are valuable not just for their material instrumentality, but also for their aesthetics. This is what we can define as the aesthetics of work proper to its productive dimension. In a certain sense, the attractive materiality of a product or the style of services depends on a material dimension of work that is also the result of making the product well and with excellence. In this sense, the quality of work does not lie in its mere functionality, but also in that the product and service are harmonious and appealing.

In other words, the result of work, although functional in the instrumental sense (e.g., a pair of shoes to protect our feet), also meets an aesthetic standard in some equally functional way, which translates into a subjective gain for users and producers. In this regard, work in its productive dimension leads to achievements that are functional not only in the material sphere, but also in the psychological, cultural, and appreciative ambits of human life. Hence, we can say that an automobile is a good product not only for its instrumentality, but also because it is a work of art.

The problem with productivity, however, is its limitation in explaining the quality of personal work. Even though products are very important to qualify work as properly well-done, we must not forget that the most valuable aspect of work is workers themselves and their personal development. Thus, in order to approach work with special attention to work as a condition for personal development and thriving, work must be assessed as a practice (praxis), that is, the internal or subjective dimension of work stands out, when that dimension conceives of action's end or its teleological meaning.

Work as a practice outlines the incorporation of meaning, purpose of action, practical deliberation and decision-making into everyday tasks. The value of work, in this way, is not found just in obtaining a product, but rather in the capacity to decide in accordance with the value that this product has for the worker and for those who benefit from it (*i.e.*, a purpose). Thus, the practical value of work is found not in efficacy or efficiency, but in making use of one's deliberative capacity and self-determination to produce something and be more efficient. In this sense, a practical dimension of work can be also defined as participation. Workers who participate in their productive and increasingly efficient tasks are those who are not simply following instructions, but they have been able to deliberate and decide better ways of production by themselves and with other workers in processes of joint deliberation (MacIntyre 2016).

Yet, the rationality associated with practices is clearly not a technique, but rather a strategic thinking that, when accurate, is called prudence (Scalzo and Alford 2016). This strategic thinking in terms of produce refers to the virtue or good habit of choosing the appropriate means in varying circumstances that make it possible to obtain a good end (*telos*). Thus, we maintain that a prudent person correctly deliberates on his or her actions, making wise practical decisions. In the case of work—as in any other human activity—prudential deliberation requires autonomy, that is, workers having the possibility to decide and impact on the way they do their job.

In addition to being productive and practical, well-done work requires to be excellent in matters related to the cooperative dimension of every occupation. Also in Aristotelian terms, every form of cooperation can be understood in light of the social dimension of human life. Following Aristotle, Aquinas argues that, driven by mutual dependency, it is inherent in every person to be social. Unlike animals, humans lack many of the resources necessary to live, or even to survive. In fact, every person depends on others for what is needed to sustain life; no one can achieve subsistence alone but requires the cooperation of others. This, in turn, is why we naturally live in society (Aquinas, *De Regno*, I, 1, 2). In this sense, the human community has both social and cooperative dimensions, which are essential for the joint effort of meeting shared needs. Hence, the concept of society can be linked to that of *company*: human society is a company of people. The term *company*, *cum pannis*, refers to two important aspects: (1) being together and organized in some way, and (2) working together to satisfy common needs. This can be illustrated by the product that fulfills a basic and universal human need: bread (*pannis*) for food (Arendt 1958). In contrast, if people were to steal from each other, they would act in opposition to cooperation, and human society would cease to exist (Aquinas, *S.Th.*, II-II, q. 66, a. 1, ad. 6).

In particular, work exhibits its cooperative dimension in two main instances: the organization and the market. Firms, in their organizational aspect, refer to the coordination of cooperative efforts toward a common objective. The market, on the other hand, represents a form of cooperation between producers and consumers, where they interact through buying and selling.

Cooperation, however, does not guarantee complete participation in the common good. In fact, a person can participate in cooperative schemes in organizations or markets just from an individual interest, like, for instance, in those cases in which a person acts under mottos such as 'scratch my back and I'll scratch yours; otherwise, nothing.' In other words, cooperative work enables the existence of organization and markets, but the existence of working and commercial communities goes one step further than is called benevolence.

This idea, in fact, can be found in Adam Smith's *Wealth of Nations* (2010) when he sustains that "… is not from the benevolence of the butcher, the brewer, or the baker that we expect our dinner, but from their regard to their own interest." In this manner, the cooperative dimension refers in this case to a relationship of mere utility or even of unabashed self-interest. A cooperative dimension of work, therefore, requires an extra dimension of fellowship or benevolence where the activity aims for the good of others, not just one's own benefit.

Comradery, the virtue of benevolence in the workplace, is characterized, among other features, by valuing the freedom of others, supporting their development, consulting one another, and sharing common goals in working places. This distinction lies between merely cooperating and truly working together, which is essential for the development of communities.

In summary, well-done work is defined by integrating efficiency in productivity, thoughtful deliberation in practice, cooperation within both the organization and the market, and comradery and fellowship as it seeks the good of others under the logic of the common good.

3.1.4 *The Promise of Professionalism: Responsibility and Integrity*

Profession comes from *professio*, which is the abstract form of *professus*, meaning 'to be recognized.' Profession means 'to attest' to some form of practical knowledge. In other words, it is recognized by society before verifying a person's performance—where there is then trust in the result of one's activity.

Since the fourteenth century, the word profession has meant 'an occupation for which someone professes to have skills.' As early as the fifteenth century, it became associated with a group of people who have an occupation, likely linked to a professional guild. During the Middle Ages, acquiring the status of 'professional' depended on associations, groups, guilds, or companies that accredited a person's professional skills. Said associations thus attest to a person's proficient practice in certain fields of action—for example, a professional caricaturist is defined as such because he works in a professional organization related to that industry, that is, a magazine.

In ethical terms, being a recognized professional is not merely a matter of accredited technical knowledge and proven expertise, such as the case with recently graduated physicians, whose competence is validated by their university professors. Being a professional also entails a promise or commitment to certain practices that can ultimately benefit other members of society—i.e. to those who, at the same time, have right to be served in a correlative manner to the corresponding professional promise.

Such a dimension of professionalism in terms of a promise, while it may lead to legal disputes, particularly in cases of negligence, cannot be guaranteed by any contract or regulation as a condition for ensuring the expected outcomes of that practice. Those who benefit from professional services must primarily rely on the professional's commitment and responsibility, rather than on economic incentives or contractual obligations. Indeed, from the Latin *respons*, being responsible means 'to respond,' 'to give an answer,' or 'to commit oneself in exchange for something'—the prefix *re* means 'in return' and *spondere* comes from the Latin 'to promise', that is, guaranteeing one's actions, being reliable, retaining in turn the meaning of 'obligation.'

In this sense, being responsible, according to Finnis (2011a), It is a form of self-imposed obligation to act with freedom and authenticity—even as tools or instruments—for the sake of achieving the best possible outcomes and promoting the common good of one's communities. The latter, in fact, is argued—in the words of Finnis—because the common good is the good of individuals living together in stable communities, where they depend on one another and collaborate mutually to foster each other's well-being.[3]

[3] As Finnis explains, being an instrument for the common good in the context of professionalism does not imply mere sacrifice. In fact, no one is responsible for attending to the well-being of others more than they attend to their own honest benefit. Responsibility, therefore, is fundamentally about the common good, where the well-being of others must be considered one's proper concern, even though no one is obligated to neglect their own well-being in the process.

In terms of responsibility and obligation, professionals must uphold the highest standards of justice, meaning they must be committed to promoting the greatest well-being associated with their work. For example, a physician who applies only the minimum standard of justice would provide treatment solely for the compensation received. In contrast, a physician who acts responsibly seeks the patient's maximum well-being and applies the necessary treatment accordingly. Therefore, responsibility is about striving for maximum justice, not just meeting a minimum standard.

Moreover, fulfilling professional responsibilities requires not only a moral disposition but also technical efficiency. Without efficiency, true service to others cannot be achieved—'to serve, be useful.'

Professionalism, as a promise, is a commitment to fellow citizens, friends, and family—not only in terms of the significant effort to provide solutions and services but also as a matter of human good. These internal goods, as MacIntyre (2016) describes them, are what allow the promise to be ethically verified by society through trust.[4]

For instance, the responsibility of physicians is recognized and valued by society not only through objective results—such as the patient's recovery or preservation of life—but also because every person, using their intellectual capacities, can recognize the value of health and life as essential to the common good of society. Similarly, professional managers are acknowledged when they commit to providing meaningful employment and promoting the well-being of fellow citizens through the affordable and sustained commercialization of products and services. It is these internal goods that enable professionals to apply their technical knowledge and resources prudently. Conversely, a physician who overuses or unnecessarily exaggerates the application of medical resources, in a situation where the patient's health could have been achieved more efficiently, exemplifies the opposite approach.

Thus, for professionals, certain internal goods are virtues and ultimate ends—such as experiences, knowledge, and goals. When engaging with others, being a professional involves striving to foster these internal goods in others. An activity is deemed good when external goods (money, technology, etc.) serve internal goods. Corruption, as mentioned earlier, arises when the relationship between external and internal goods becomes distorted.

On the other hand, a lack of professionalism, in ethical terms—not just due to deficient technical knowledge or performance—manifests as professional corruption. This corruption is characterized by at least one of the following conditions: (1) acting for instrumental goods (money, power, status) as though they were internal goods, and (2) making decisions based on individualistic motivations rather than aiming to cooperate for the common good of the community.

To counter these risks of corruption, professionalism must be reinforced by the practice of integrity, which involves acting in pursuit of both internal and common goods. In this context, an action is characterized by integrity not only because it

[4] The English word 'trust' has its etymological origins in the medieval Viking *traust*, meaning help, confidence, protection, or support. In this sense, 'trust' is connected to actions that generate trustworthiness, rather than to the act of making a promise. Responsibility, in this context, is assumed. Trust, however, is granted not only through the explicit recognition of responsibility, but also through the actions that demonstrate a professional's ability to be trusted.

connects instrumental goods to ultimate goods, but because it places the former in service of the latter. Therefore, simply renouncing instrumental goods like money, power, or prestige does not define integrity. Rather, integrity is characterized by the integration and proper ordering of both instrumental and final goods, ensuring the superior importance of the latter.

Within this framework of integration, Finnis (2011a) explains that medieval philosophy adhered to the maxim *bonum ex integra causa, malum ex quocumque defectu*, meaning an act is morally good when all that it entails is good. Conversely, an act is morally bad if it is defective either in how it orders instrumental goods or in how it seeks these goods for their own sake.

This principle can be applied to a physician's integral action, which involves seeking the health of patients within the context of clinical and institutional efficiency. In other words, a professional's integral performance must take institutional efficiency into account, recognizing that ultimate and superior goods transcend mere instrumental ones.

Finally, truly integral action inherently involves the common good. This means that one's personal good and development must be integrated with the well-being of others in the communities to which one belongs. This represents a form of personal-community integrity, where there are no zero-sum games and no actions taken at the expense of others.

3.1.5 Practices Associated with Professionalism

As we will show, *we* can be considered a non-exhaustive list of practices applicable to all kinds of professionals, regardless of the position or role they currently hold. Accordingly, we will consider several practices such as compliance, accountability, conflicts of interest, personal behavior, professional objectivity, creative accounting, and employees' loyalty, as follows:

3.1.5.1 Compliance

This practice refers to the habitual practice of adhering to standards, whether related to legislation or the company's internal codes of conduct.

When there are irregularities, taking distance from the irregularity is required, as is immediate cooperation to amend and rectify it. In order to act in accordance with the rules, an attitude of compliance requires, in turn, the responsibility of staying on top of regulations. Otherwise, negligent misconduct can arise.

As for the company, it must provide access to regulatory information, instructing company members on the administrative processes related to compliance or non-compliance with regulations, without forgetting to provide access to legal advice when the circumstances warrant it.

3.1.5.2 Accountability

This one is the practice characterized by performing verifiable and transparent managerial actions. Verification is justified by the need to support one's professional decisions over time.

Thus, being accountable implies not only having the correct intentions (being honest) but also being able to justify them and eventually being able to demonstrate that one has acted in good faith.

In this sense, transparency is valuable when it refers to the possibility of concretely clarifying intentionality in the sphere of professional decision-making. Thus, it is always possible to attest to and provide evidence for what one has done and the intention behind one's own actions.

3.1.5.3 Disclosing Conflicts of Interest

In general terms, a conflict of interest refers to circumstances in which a person in a given role (e.g., a physician) cannot guarantee that their decisions are entirely free from compromise—that is, that their objectivity and impartiality have not been, and will not be, affected. Such objectivity relates directly to what is expected from the role itself, namely, the purposes that define it and guide its decisions. For example, in the case of a physician, these purposes include the health and recovery of patients. In this sense, a conflict of interest is fundamentally explained in relation to the concepts of purpose and objectivity in the decisions of a person in a given role, as well as the existence of third parties with a legitimate interest in demanding objectivity and impartiality from those who perform that role.

Thompson (2017) explains that a conflict of interest arises when professional judgment regarding a primary goal (such as a patient's well-being or the validity of research) could be unduly influenced by a secondary interest (such as financial gain). In such circumstances, the objectivity of the decision cannot be reliably verified by a third party, and there is no assurance that the professional has acted in an ethically sound manner.

The usual practice for addressing situations involving a conflict of interest is known as disclosure, which refers to making existing conflict of interest public or known to the corresponding stakeholders. In sum, when identified, conflicts of interest should be openly declared and effectively communicated so that corresponding third parties can assess whether there is a risk of bias.

3.1.5.4 Professional Behavior

Professional behavior involves criteria of honorability, integrity, reputation, and competence. It requires maintaining a safe distance from activities that are dishonest, dubious, or based on lies, etc. This also applies to the private sphere as employees' private activity can impact the prestige of the company and one's own professional reputation.

3.1.5.5 Professional Objectivity

This practice means identifying conflicts of interest in the company and/or among clients means maintaining a balanced relationship between the interests of customers and the company, without favoring one party at the expense of the other. This implies the need to prohibit or regulate favors and gifts from customers and productivity bonuses from the company.

3.1.5.6 Creative Accounting

This refers to the fraudulent manipulation of accounting and financial information ('cooking the books'), in such a way that there is either a direct (personal) or indirect (for the company) benefit.

3.1.5.7 Employee Loyalty

This refers to the need to safeguard company interests, without subordinating them to personal obligations. In this respect, loyalty to former employers also applies, avoiding, for example, disclosing privileged information, discrediting the company with clients before leaving the job, or appropriating and transferring clients or databases to new employers.

3.2 Working Conditions: Good Work as a Dignifying Activity

The circumstances in which any human work is performed must be considered due to their impact on personal development. In ethical terms, these circumstances are commonly referred to as working conditions—the factors that either support or hinder workers' personal growth in the workplace.

The idea of working conditions as impacting the quality of work was firstly proposed by the Australian organizational psychologist Elton Mayo. However, since his *Hawthorne Studies* the consideration of different conditions of work has gained importance due to their effects on workers' productivity. In other words, improving working conditions has turned into a way to improve productivity.

From an ethical standpoint, this argument, while not incorrect, is quite reductionist, as it risks becoming a justification for attending working conditions mainly for the sake of sustaining or improving productivity, not for workers themselves. In a different manner—clearly less utilitarian—the value of those features that constitute the circumstances in which work is performed are a common good and, therefore, they must be oriented firstly to the personal development of those same workers, which is, according to a virtue ethics approach, a matter of dignity.

3.2.1 The Principle of Dignity

To explain the dignity of work, three considerations must be addressed: (1) what dignity means, (2) how it applies to human work, and (3) what it requires once recognized.

Regarding the first consideration, the term *dignity* derives etymologically from the Latin *dignus* (nominative *dignitas*), meaning 'valuable.' According to the Thomistic tradition, this principle refers to the value of every human being as a person—namely, the dignity of the human person—which is valuable in itself, not because of its utility or usefulness, that is, not in relation to something else (Aquinas, *Commentary on the Sentences*, III, d. 35, q. 1, a.). Rather, dignity refers to what is valuable in an inherent, ontological, or intrinsic sense.

Now, such intrinsic value must be considered in a twofold—and complementary—manner, namely, in static or dynamic terms.

According to de Souza Salles et al. (2022), human dignity—understood in a static sense—means that the reality of the existing person has intrinsic value, both materially and subjectively. This can be argued in two ways. First, in comparison with other beings, because our rational nature is more valuable than animal or vegetative life. In this regard, each person has value, but that value, unlike other living beings, is granted a superior role because each person has an intrinsic value that is evidenced through exercise of the will and intelligence. Second, personal life possesses inherent worth as it is both unique (entirely original) and integral. As for the first, we can sustain that, while all human persons share the same nature, each one lives in an individual and unrepeatable way. Accordingly, we can say that each person is, more precisely, someone—a singular and fully original being. As for the second, this unique mode of existence is expressed integrally, meaning that it is contrary to the personal way of living to act or treat others solely according to certain dimensions of a personal existence—in other words, in a reductionist manner. It violates the dignity of the person to consider oneself or treat others solely in material (instrumental) terms, or to value only one's own or others' subjective dimension. Every person must be treated in accordance with their integral nature.

> The individual who is a person cannot communicate with other persons as a part, because he is a complete whole. Moreover, the person cannot communicate as the universal communicates to particulars, because the person is something subsistent in itself. Finally, the person cannot communicate as something assumable by another, because there is not more than one personality of its own, and what is assumable is based on the personality of the one who assumes. (Aquinas, *Commentary on the Sentences*, III, d. 5, a. 2, a. 1, ad 2)

On the other hand, personal dignity must be considered in a dynamic manner—that is, as the capacity to be the cause of one's own actions and, consequently, of one's personal fulfillment. This is, in other words, the idea of moral dignity, which is distinct from ontological dignity.

As de Souza Salles et al. (2022) state, this dynamic—moral—dignity is the dignity that a persons can tend to by his own freedom and voluntary acts. Is, in a way of speaking, a dignity to be achieved through the dominion of their personal action—acting by themself—and, according to this, there are differences between each person, when, as explained by Aquinas, acting out of our own freedom is more worthy than not acting freely (Aquinas, *De Potentia*, q. 1, a. 5, co.). This form of dignity, therefore, is achieved according to our own capacity to master our own actions to do good, and not being moved by others (Aquinas, *S.Th.*, I, q. 29, a. 1).

This is, in the words of Finnis (2011a), the dignity of self-constitution according to which individuals can value themselves as being responsible agents, that is, not living their lives for the convenience of others, neither acting themselves nor acknowledging others just for instrumental purposes. According to this, each person has a moral dignity based on the radical capacity to deliberate and choose, for inner and outer determination, expressed as the

> self-determining creativity for participating in intelligible goods. Personal life accordingly has the dignity which the tradition sought to capture with the phrase 'image of God'—a phrase which serious philosophers such as Socrates, Plato, and Aristotle would not have dismissed as a mere theological flourish foreign to philosophy's reflection on the ultimate principles of everything. (Finnis, 2011b, p. 219)

Regarding the implications of personal dignity for human work, perhaps the most significant consideration stems from the fact that each person is entitled to deliberate and choose, possessing the autonomy necessary for self-determination in relation to what the productive demands of regular tasks and duties can offer (Pinto-Garay et al. 2020). As MacIntyre's neo-Aristotelianism (2016) suggests, working conditions should be regarded as common goods shared by those who work together—a common good achieved by producing goods and services that contribute to the life of the community, and by doing so in a way that enables those same workers to become excellent at producing them through the exercise of their deliberative capacities.

Therefore, no worker has value as a mere instrument, that is, no working person can ever be valued for their usefulness, unless their activity, in achieving 'useful' results, is at the same time a condition for personal development through the exercise of her will and understanding.

As for what is required for each of us once the dignity of others is acknowledged, we can say, in line with the above, that, according to the dynamic or moral value of each person, we should be impelled to act with care. In short, dignity demands care. Indeed, the basis of human dignity lies in recognizing that human development and perfection have no limit, which is due to the fact that each person's nature has incalculable potential. Therefore, the dignity of each person leads to a type of personal relationship that manifests itself in acts of care (from the Latin *cogitatus*) that imply permanent attention to personal development and well-being. Caring, in this sense, means understanding and tending to the dignity of people in such a way as to protect their development integrally, from the most basic or physical aspects to the more subtle dimensions of human living.

Now, in terms of development, we must consider the conditions that allow for greater degrees of freedom and reflection on the part of people, including promoting decision-making and their reflective capacity. In terms of well-being, this means the material conditions that are evidence of a good life, from the most hygienic to the most aesthetic elements.

Thus, personal dignity establishes the meaning and justification of all social, economic, and political activity. Yet, society, economics and politics are for each human person, not, by contrast, each human person for them. This means that everything carried out by public or private institutions, organizations, and the market must be oriented toward the development and well-being of people. In the same way, no person who participates in the social, political, and economic spheres can be a mere instrument for others' development and welfare (like with slavery). Rather, everyone presents an opportunity for others' development and welfare in a relationship based on participation, not on utility. Hence, people are never merely a resource for a broader general end but are always an end in themselves.

Personal dignity, therefore, establishes two important practical criteria for economic and business relations. The first is a more restrictive criterion, according to which a person should never be mistreated, regardless of their merits, actions, education, or abilities (in this context, mistreatment is understood as the opposite of care). The second criterion is more proactive, asserting that personal relationships should foster development and well-being. In other words, one should, to the extent possible, strive to improve the lives of those with whom one has a work-related or economic relationship. This, indeed, is consistent with the fact that many companies ultimately aim to promote the development and welfare of their workers and customers.

3.2.2 *Practices Associated with Working Conditions*

Derived from the notion of dignity and the corresponding responsibility to 'care' for those who participate in the firm as employees, it is important to examine in which aspects a person's work must be taken into care. Accordingly, two fundamental considerations arise: first, how the so-called working conditions are managed; and second, what kinds of work activities can never be morally accepted, regardless of the conditions under which they are performed.

3.2.2.1 Managing Working Conditions

In the first place, we can consider physical working conditions, that is, the material aspects of a workplace related to the physical well-being of workers. This dimension encompasses elements such as cafeterias, restrooms, safety equipment, breaks, and so on. The aesthetic conditions of the workplace also play a significant role in this regard.

Unethical working conditions, in this context, include what are commonly known as sweatshops. Originally, the term referred to the garment industry during the first Industrial Revolution in the nineteenth century. It is associated with extremely harsh physical work conditions, hence the use of the word 'sweat.' These jobs are strenuous due to the physical effort required, not necessarily because of task repetition. Activities that fall under this category include construction, mining, night shifts, military service, or high-performance sports. These jobs not only lead to exhaustion but also cause long-term damage to health and life expectancy, as seen in cases like mining-related silicosis.

Psychological working conditions focus on how workers are treated and how that treatment affects or protects their mental health. Psychological conditions should be managed to prevent issues like stress, burnout, or bullying, among other negative practices.

A well-known example of unethical psychological working conditions is mobbing. This practice refers to mistreatment in the workplace and is often described as workplace harassment or bullying aimed at provoking resignation or simply as a form of inappropriate treatment. Mobbing involves not only psychological mistreatment (bullying) but also the imposition of unattainable goals and objectives, which can lead employees to resign due to exhaustion. As such, mobbing is not only a violation of psychological and legal standards but is also a sign of poor management of productive resources and time.

Axiological working conditions pertain to the values or principles upheld by both companies and individuals. Etymologically, the term comes from the Greek *axios* (ἄξιος), meaning 'value.' Axiology is the study of values or principles present in society and its institutions. These values can be moral, religious, cultural, political, and so on. Axiological conditions are therefore concerned with respecting the personal values of an organization's members

Social working conditions refer to the need for individuals to interact with others. In the context of companies, this interaction includes not only professional contacts but also the social life within the organization. Thus, social working conditions also encompass informal aspects that foster camaraderie. Within this category, there are also conditions related to organizational justice, which involve the structures of relationships between people or compensation policies that fail to justly recognize the merits or rights of some workers, while doing so for others.

Legal working conditions pertain to the contractual aspects that govern labor in most of the world's markets. This includes formal contracts as well as the adherence to duties and rights that both workers and employers must uphold. Informal work, for instance, represents a form of non-compliance with this condition.

Another crucial aspect of working conditions concerns the availability of resources for work and occupational health. Related to psychological well-being, a lack of adequate tools or resources not only reduces productivity but can also negatively affect the worker, causing habitual frustration due to unsatisfactory results. Therefore, working conditions related to resources involve aligning demands, development, and goals with the available means to accomplish tasks.

In terms of time, working conditions refer to the compatibility of work schedules with lifestyles that prioritize rest, family and social life, health, and education. This includes the amount of time work occupies in a worker's day or week—ranging from commuting times and restrictions on 'in-office' activities, to the need for disconnection from work-related tasks on digital platforms (such as the right to disconnect).

3.2.2.2 Restricting Illicit Labor

From the perspective of business ethics, illicit jobs refer to those that, regardless of the conditions under which they are performed, are always unacceptable in terms of human dignity, even when they are performed freely. Such jobs, therefore, are always ethically illicit because they involve the objective degradation of those who perform them.

Now, there are different types of illicit labor, and although they are different in their circumstances and forms, all of them fail to respect personal dignity in equal measures. Among others, we can consider the following:

Child labor is illegal for two reasons: First, because of the consideration of an alternative cost. In the case of children, it is better for them to play or study rather than working. Second, child labor does not refer to a sporadic occupation, but to a real job, with all the demands and responsibilities that it entails. Therefore, this work implies a type of activity that, because of its workload, harms children. In more concrete terms, as a real job, child labor causes physical and psychological damage. Development is not only stopped or impeded, but is actually prevented, and this is a serious human dignity problem.

Forced labor, if only associated with slavery, tends to be a limited assessment. There are other forms of forced labor, such as that performed by prisoners and conscript labor.

Regarding slavery, it is important to consider that the most serious form of slavery on record is modern slavery, especially in the antebellum southern United States. This is not just because of the arbitrary restriction of freedoms imposed there, but also because of the abuse and mistreatment. Now, although slavery has been abolished, it is still a social phenomenon evidenced in the trafficking of immigrants or in sex trafficking.

As for armies that still have compulsory conscription (such as Israel or South Korea), the matter is different because compulsory service is not subject to a productive purpose. In the case of compulsory military service, the justification is national defense, which is an essential public good.

In the case of prison systems in which prisoners work, it is possible to argue that their work is licit since, once the working conditions are safeguarded, the tasks assigned in prison can be justified in terms of social insertion and the employability of prisoners. In other words, it resolves factors that tend to encourage crime.

On the other hand, prostitution or sex work is not necessarily forced labor, although the prostitution industry does include this type of practice (sex trafficking). However, in the case of sex workers who perform this activity freely—even under conditions of social acceptance and legality—there is no way to consider this work as licit. Strictly speaking, the fact that an activity is carried out freely does not mean that it is good, even though a good action must necessarily be free. Nor is it true that everything that is legal, or is at least not illegal, is morally correct. Thus, offering and consuming this type of service is illicit because it is harmful. In other words, it deteriorates good living in physical, psychological, and biographical terms. At the same time, it involves the logic of instrumentalization. There is no sense of donation or intimacy, thus degrading the human value of sexuality by turning the sex worker into a simple good for consumption.

Also considered an ethically illicit form of labor, criminality refers, in this case, to offenses, misdemeanors, or felonies. Although they may be typified in norms or laws, in the deepest sense, crimes not only disobey established norms but also and specially produce social harm. Strictly speaking, the law effectively contains a description of actions that are harmful to the public good. Therefore, committing a crime is essentially an attack on the common good, both because it damages it and because it impedes its development. Criminal work, therefore, is not simply acting against the law, but rather harming the social order. Hence, the activities of drug traffickers or hired assassins are illicit from a moral point of view, regardless of the conditions under which the related jobs are carried out.

3.3 Participation and Productivity

The organization of work—particularly industrial work—is generally understood as the application of systems of specialization and division of labor aimed at increasing efficiency and productivity. However, this practice—also known as work design—has not been viewed solely as the application of technical criteria to organized work. From the 18th century to the present, it has sparked a discussion about the effects on the workers themselves who participate in these industrial processes in terms of trying to understand how the increasing productivity would affect or not workers personal development.

This problem was originally described by Adam Smith and opened the path for the early discussion of what are now referred to as ethical issues in work organizations (Pinto et al. 2022). According to Smith, the division of labor, which is based on an efficient arrangement of simple tasks, increases productivity, even with uneducated employees:

> ...the important business of making a pin is, in this manner, divided into about eighteen distinct operations...though in others the same man will sometimes perform two or three of them... But if they had all wrought separately and independently, and without any of them having been educated to this peculiar business, they certainly could not each of them have made twenty, perhaps not one pin in a day; ... part of what they are at present capable of performing,

> in consequence of a proper division and combination of their different operations. (Smith, WN, book I, chapter I, article I).

In this context, Smith suggests that the division of labor can offer workers an opportunity of personal development:

> The greatest improvement in the productive powers of labor, and the greater part of the skill, dexterity, and judgment with which it is anywhere directed, or applied, seem to have been the effects of the division of labor (…). (Smith, WN, book I, chapter I, article I)

However, the same organizational system of production, in certain cases, causes workers to deteriorate in their rational capacity:

> But the understandings of the greater part of men are necessarily formed by their ordinary employments. The man whose whole life is spent in performing a few simple operations … has no occasion to exert his understanding or to exercise his invention in finding out expedients for removing difficulties which never occur. He naturally loses, therefore, the habit of such exertion, and generally becomes as stupid and ignorant as it is possible for a human creature to become. (Smith, WN, book V, chapter I, article III).

Therefore, the division of labor, to the extent that it provides specialization and autonomous spaces for inventiveness, can be an opportunity for development, although this is not guaranteed by the mere division of labor. Smith does not claim that specialized work necessarily produces a degradation of workers or their moral development; both possibilities remain open (Rosenberg 1965). Claiming that the division of labor necessarily causes harm to workers corresponds to a Marxist view of productivity and labor.

In a different manner, the neo-Aristotelian approach to productive organizations managed under the logic of specialization and the division of work, indicates that some considerations must be made to keep productive tasks as a condition for personal thriving. In this vein, MacIntyre (2016) sustains that any productive work can be virtuous and excellent only if it is performed in a deliberative manner, like in practices (*praxis*), but this is possible only under the possibility for workers to participate—not just operate—in production (Pinto-Garay et al. 2022).

3.3.1 Work Design and Living Conditions

Job design refers to the principles that guide the planning of tasks, responsibilities, and coordination policies among the various members of a company, society, the market, and the workforce at large. These principles emerged alongside the industrialization of North America in the twentieth century, with prominent figures such as Henry Ford and Frederick W. Taylor leading their development. In general terms, the 'ethical' considerations in these systems were not excluded but were formulated differently, separate from the logic of praxis and participation. Early job design models in the twentieth century focused on increasing productivity through the integration of specialization and division of labor, while also aiming to improve workers' social conditions and material welfare—though without incorporating any form of participatory schemes for the workers themselves.

In this regard, twentieth century job design—particularly through the ideas of Henry Ford and Frederick Winslow Taylor—introduced the concept of social welfare as a development criterion for employees, but linked to the rise of productivity, in a way in which employment and productivity were reformulated in terms of lifestyle and consumption.

Henry Ford (1863–1947), an American businessman and founder of the Ford Motor Company, was a pioneer of modern industrialization, introducing assembly lines and stimulating consumption in the United States as part of a broader model of social welfare. His ideas on work organization were implemented in conjunction with his vision of the country's social and political progress.

Ford built his company on several key principles: (1) specialization through task breakdown; (2) a general division of the company based on production, assembly, and administration (vertical integration); (3) viewing each employee, skilled and unskilled, as part of a large production machine ('a machine made of men'); (4) decision-making based on technological flexibility, allowing for the manufacture of components and parts usable in various products; and (5) seeing employees as both workers and consumers. As a result of these principles, Ford believed that increased productivity would have a double effect: higher wages and lower manufacturing costs. This, in turn, would grant workers access to a market previously reserved for the upper classes. Ford's ideas led to a significant boost in productivity. He implemented anti-absenteeism and turnover measures, reducing the respective rates from 10% to 0.5% and from 400% to 15%. These results were achieved through a company-wide wage increase and a profit-sharing scheme—policies that were groundbreaking at the time.

All of this was paired with Ford's definition of the 'New Man,' an ideal employee who was not only a worker but also a citizen and consumer. Ford's concept of the New Man was justified not only by the resulting increase in productivity but also by the creation of new markets. By making automobile production more efficient, Ford aimed to raise wages, thereby integrating the working class into markets that had traditionally been aimed at wealthier citizens, such as the emerging automobile market. In broad terms, more productive workers would be able to purchase the very products they helped manufacture. This would give rise to a new social class—the middle class—characterized by higher living standards and greater consumer power. Thus, Ford's vision of social progress became intertwined with the growth of jobs and mass consumption. Ford's New Man would not only embody certain normative or moral traits but would also be able to enjoy a high standard of living.

On the other hand, Frederick Winslow Taylor (1856–1915), an American engineer, is considered the father of modern management. As the author of *The Principles of Scientific Management* (1911), his principles for organizing work were widely implemented in twentieth-century industry. In *The Principles of Scientific Management*, Taylor (1911) not only presents a series of technical principles aimed at improving organizational efficiency through the reorganization of tasks but also incorporates an ideological element by seeking to enhance the welfare and quality of life for workers. Taylor combined the need to improve workers' income with the goal of increasing productivity. Although Taylor genuinely cared about the welfare of his workers, he regarded them as somewhat crude (he initially used the term 'stupidity' to describe them) and believed they were only capable of assuming a passive role as operators. In other words, he thought they lacked initiative and should follow instructions quickly and without question. According to Taylor, this perspective led to the need for a separation between planning and execution within an organization. Thus, Taylor's model relied on limiting workers' autonomy, deliberation, and practical judgment.

By constraining deliberation, Taylor aimed to increase productivity, replacing it with 'scientific' organizational laws and techniques that could be replicated across different enterprises. Productive methods were seen as large machines, with workers serving merely as instrumental components in the process (Breen 2012). In this way, the Taylorist company becomes a highly efficient organization based on centralized scientific or technical 'organizational knowledge.' This knowledge does not refer to the skills of the workers themselves but to company-defined knowledge that dictates employees' tasks. To be developed and executed, this knowledge is concentrated in the planning department at the heart of the company. Taylor's approach to increased productivity was grounded in the study of workers' movements (ergometry). This study aimed to make work more efficient by going beyond the division of labor to consider the individual characteristics of workers. In doing so, Taylor sought to enhance the efficiency of both organizations and their workers.

As we have seen, both Ford and Taylor recognized the importance of increasing productivity in part through improved living standards—an essential consideration. However, they did so in a way that left workers with almost no autonomy, thereby limiting the practical dimension of work, as previously discussed. In this sense, an essential ethical dimension of work was neglected, a critique particularly leveled against Taylorism by many scholars. Thus, the challenge was for management theory and business ethics to reintroduce workers autonomy to enhance productivity.

3.3.2 Autonomy and Participation

During the last decades of the twentieth century, the importance of autonomy and workers' participation emerged from two key perspectives: management theory and business ethics. Often referred to as participatory management, this approach is grounded in the principle that employing a worker involves not only integrating them into the productive system through active participation but also leveraging their motivation and capabilities to enhance the company's benefits (positive outcomes).

In moral terms, the management principle of participation addresses the fact that employing people does not signify the same as using them—If this were so, all kinds of employment would be morally unacceptable as being only instrumental to the firm, that is, disregarding that people can never be considered just as means, but rather ends in themselves.

Hence, to avoid the instrumentalization of workers, organizations must design tasks according to the principle of involving employees not only in material and objective terms but also subjectively. In other words, this means giving employees the capacity for self-determination within an autonomous context that allows them to perform their work with genuine spaces for deliberation (the practical dimension of work). This distinction enables us to view the integration of workers as fundamentally different from their instrumentalization; it fosters the development of a practical dimension of work (*praxis*), rather than merely a productive one (*poiesis*). Only in this way can workers be integrated into companies in a participatory manner, rather than being used (instrumentalized) by them. The results of individual tasks may have an instrumental nature, but the workers themselves do not.

As stated before (see introduction), MacIntyre (2016), following Aristotle, argues that such an organization, based on autonomy and personal deliberation, can be seen in certain Japanese industries, where workers are integrated as team members. In these teams, each assumes responsibility for the entire production process involved in manufacturing an automobile. Thus, good or excellent products result in particular from deliberative work or, more specifically, work performed in terms of a practice.

According to MacIntyre, practices are complex social activities that pursue goods or objectives internal to the activities themselves, much like Aristotelian praxis (MacIntyre 1988). A practice, henceforth, is worthwhile and valuable *per se* because the agent gains something in the very act of performing them. The purpose of *praxis*, therefore, is to contribute to the agent's personal or moral development and, consequently, to foster the acquisition of experience and habits; indeed, the agent herself becomes the ultimate purpose of the activity and its most valuable outcome (Sison et al. 2016). MacIntyre calls such outcomes or goods internal to distinguish them from instrumental or external goods associated with productive activities (MacIntyre 1981). According to MacIntyre (1988, p. 32), external goods can also be described as goods of effectiveness, while internal goods are goods of excellence.

For MacIntyre, one of the main reasons for recognizing human action as a practice aimed at excellence is that it is performed on the basis of the habit of practical rationality and deliberation. For him, the most significant virtue for practice is prudence—practical wisdom—or the good habit of sound practical judgment (2016). Therefore, whenever work incorporates good practical rationality, it can be considered open to virtue and excellence. In Aristotle's view, all human activity is potentially perfective when exercised within the context of rationality (Nagel 1972).

Specifically, practice entails the capacity to exercise practical rationality in selecting the means best suited to achieving a good end (*telos*), which corresponds to the central virtue for acting, *phronesis* or practical wisdom. Without this intellectual virtue, other virtues cannot be exercised, and excellence remains unattainable (MacIntyre 1981). *Phronesis* results in the power to understand and respond to intelligible goods, and to choose and implement one's decisions well. This perfection involves harmonizing the powers of intellect, will, and emotion both with one another and with the human good (Finnis 1998). For Aristotle, *phronesis* or practical wisdom is not merely the ability to choose the means to an end but also includes a moral disposition toward a known good that benefits the agent or another person (Aristotle, NE 1140b), that is, in terms of a common good. Moreover, the agent also knows how to realize that good by deciding on the particular means for achieving it (MacIntyre

1985). In the same vein, Aquinas explained that practical wisdom, or *prudentia*, can be understood as the practical knowledge of what is to be done (Aquinas, *S.Th.* II–II, q.47, a.4) when deliberating on and selecting the means appropriate to a given end (Aquinas, *S.Th.*, II–II, q.47, a.6). In sum, work qualifies as a practice insofar as it is carried out in accordance with practical rationality and performed in a deliberative manner; only in this sense is it open to virtue and excellence. This means that work, as a potentially perfective action, must integrate both (1) practical reasoning (*logos*) in production (*poiesis*) and (2) a connection between personal excellence (*areté*) (Gómez-Lobo 1989) and the common good (Sison and Fontrodona 2012, 2013).

In this context, autonomy at work represents a positive organizational feature characterized by the thoughtful execution of tasks that involve applying initiative within the company. Autonomous work requires the company to provide adequate space and resources for deliberation and self-determination, enabling employees to pursue their own objectives (aligned with the company's goals) within a framework of task optimization (Sayer 2009). In this regard, companies must empower their workers, which not only enhances company performance but also benefits the employees themselves, helping them avoid the experience of alienation at work—such as stress, frustration, apathy, and anxiety.

In the history of management, this type of work design based on autonomy and self-determination of employees was originally referred to as 'Toyotism,' as it was first implemented by Toyota in the 1970s. Toyota was, in fact, the first to encourage operators to participate in the design and improvement of industrial processes to enhance process efficiency, even when these operators were not engineers or specialized personnel. Later, this organizational model, also known as Quality Circles, was further refined with other production strategies, such as Lean Manufacturing, Total Quality Management (TQM), and Just-in-Time Production—systems that rely on the active participation of all those directly involved in the production process, including even the most basic operators.

Also, in line with Aristotle's ethics, the Japanese author Nonaka et al. (2008) argues that reformulating management theory based on Aristotelian ethics opens a better understanding of action within organizations as the union of practical knowledge and production. In this sense, it is quite important that autonomy and self-determination not only must be extended to all employees and done as in teamwork, but it also needs to establish some boundaries in terms of objectives, responsibilities, and authority.

As extended self-determination, an Aristotelian-based theory of work design asserts that autonomy must be granted to both managers and non-management employees. Therefore, all members of the company should have the opportunity to make decisions related to their own tasks, based on their decision-making and technical skills. On the other hand, shared deliberation for personal self-determination means that deliberative processes within the company must be carried out collaboratively. Final decisions should not be made by an individual employee alone, but rather by work teams acting together.

In terms of boundaries, autonomy for self-determination must be established under productive terms. This means that all spaces for autonomy and self-determination at work must be directed toward what primarily, though not exclusively, justifies a position within the organization in productive terms—namely, efficacy and efficiency. This means that self-determination must focus on improving the productive quality of work. Granting autonomy does not imply the right to reflect on or offer opinions about all company matters, but only on those areas that the organization has defined as part of the specific responsibilities associated with a job description.

Additionally, autonomy must be defined in its boundaries considering the organization´s need for authority. workers at the bottom of the company's hierarchical ladder usually deliberate on aspects that are less important for the overall functioning of the company. In this sense, the autonomy of managers implies different standards of responsibility with the firm's success.

According to said autonomy boundaries—productivity and authority—we can argue that participatory management is something different from organizational democracy (OD, henceforth), in which the limits of self-determination are not quite clear, particularly those associated with the common good of the firm. OD aims to increase participation within the firm by introducing a political-activist dimension to work design, while still emphasizing productivity and other positive organizational outcomes. According to Harrison and Freeman (2004), individuals within OD appreciate the opportunity to be heard in organizations or to have some influence over decisions. As a result, OD can enhance employees' sense of belonging and the meaningfulness of their work. Participation in decision-making tends to foster greater commitment to the final decisions, which can facilitate their implementation. OD also instils a sense of greater responsibility for the company's outcomes, potentially reducing actions that conflict with social values when employees view them as their own. Furthermore, OD promotes the development of more participatory environments, which can lead to higher innovation rates and a greater willingness to embrace change. Within a more democratic organization, participation helps employees develop decision-making skills.

Despite the benefits mentioned, OD often presents challenges related to technical limitations (Kerr 2004). For example, when employees at the operational level are involved in decision-making, it may lead to decisions that harm the organization due to factors such as a lack of education, professional training, or necessary experience. Furthermore, extensive democratic processes can be time-consuming, which undermines efficiency, especially when decision-making relies on negotiation processes without pre-established limits. Democratic processes can also encourage an excessive tendency toward organizational change, potentially disrupting effective systems already in place.

On the other hand, democratic processes may reduce management's authority or result in increased audits and accountability measures, which employees might eventually reject. Additionally, OD may not align with the company's need to make quick decisions on certain strategic matters—decisions that are typically made by those in control. Most importantly, democratic decision-making does not always lead to the best ethical outcomes, such as improving operational performance and ensuring the benefits for shareholders, suppliers, or consumers.

In general terms, such organizational risks derived from OD can be understood as the result of treating participation as a goal, rather than as a means of fostering self-development for the greater good of the firm. In this regard, while participation and autonomy are essential for assessing work design in ethical terms, they should never be seen as the ultimate objective in ethical terms. Instead, they must be considered as necessary conditions for achieving higher standards of virtue and excellence within the firm.

3.3.3 *Practices Associated with Employment and Participation*

In light of the principle of participation, several practices must be considered not only in terms of what we have described as deliberation—under which we might include practices such as the implementation of Quality Circles or Lean Manufacturing processes—but also in terms of participation conditions not directly related to autonomy and deliberation as we will show below:

3.3.3.1 Discrimination

To discriminate, in its literal sense, means to recognize and understand differences, and based on this recognition, to choose or exclude. This action is not inherently negative. However, when discrimination occurs in the context of accepting or rejecting people—particularly when it is based on arbitrary reasons—it becomes unjust.

The International Labour Organization (ILO) (1958) defines labor discrimination as the distinction, exclusion, or preferential treatment based on race, color, sex, religion, union membership, political opinion, or any other irrational or unjustified motive. Such discrimination alters or nullifies equal treatment in employment and occupation.

Discrimination can therefore be understood in two ways: The first is negative discrimination, which involves decisions or actions based on arbitrary criteria. This form of discrimination is unjustified and unfair from legal, practical, or moral standpoints.

The second type aligns with the literal definition of discrimination. In this case, decisions are made based on criteria that are justified by law, practical considerations, or fundamental rights. While this may be referred to as discrimination, it represents a just form of selection because it is based on known and accepted criteria.

The problem with arbitrary discrimination is not just the absence of justification, but also the presence of unacceptable justifications. Such decisions harm individuals by impeding their development or failing to treat them with dignity. This issue arises not only from a lack of practical rationality (ignoring legal or practical principles) but also from a failure to uphold justice.

There are generally three types of criteria used to assess whether a decision is just or arbitrary. The first is the legal criterion, which requires companies to comply with existing regulations. The second is the functional criterion, which relates an organization's technical or productive requirements to specific roles and responsibilities. For example, selecting a candidate based on technical competencies aligns with this criterion. Finally, there is the criterion associated with the public good, which emphasizes a firm's responsibility toward justice, the common good, personal dignity, and human rights—especially when no specific legislation exists to guide decision-making.

Arbitrary discrimination in the workplace can take two forms: direct and indirect. Direct discrimination occurs when a decision is made for explicit and clear reasons. Indirect discrimination, on the other hand, involves decisions made with seemingly reasonable justifications, but these reasons are implicitly arbitrary.

Discrimination is not limited to hiring or dismissal processes. It can also manifest in career development, such as when an employee's professional growth is stunted, or they are denied opportunities without justified or reasonable cause.

Managing Termination or Firing

Termination is an inherently ethical matter. The decision to fire an employee can be likened to a sentence—a decision that is often difficult for both the person making the decision and the person being terminated.

In our opinion, termination should not be seen as a regular aspect of business operations. An employee is hired not to be dismissed, but to contribute and potentially grow within the company. Therefore, termination must first and foremost be viewed as an exceptional event.

However, dismissal is sometimes unavoidable. Even though it may be wrong in many respects—whether due to legal, economic, or organizational reasons—it is not always immediately clear which normative principles should guide the decision.

From the perspective of contributive-distributive justice, employee termination may be justified when an employee's contribution is insufficient, or when distribution (of work or resources) is no longer feasible for the company.

Accordingly, employees' contributions can be understood in several key ways: technical, decisional, cooperative, and institutional.

Technical contribution refers to the skills and knowledge employees bring, which directly support productivity. A worker's technical expertise translates into tangible output and the achievement of productivity standards.

Decisional contribution recognizes that nearly all roles involve decision-making. Whether in autonomous workspaces or as part of a team, the ability to make sound decisions is integral to an employee's value.

Cooperative contribution involves the ability to work effectively with others—superiors, peers, or subordinates. This is not just a matter of organizational skills but also of character. Much of an organization's success depends on qualities such as communication, clarity, understanding, loyalty, and a willingness to collaborate.

Institutional contribution refers to the extent to which employees align with and support the values and principles upheld by the company, such as its mission, vision, and culture. For example, companies like Nike and Reebok prioritize health and sports. An employee who views these pursuits as a waste of time would have a limited institutional contribution in such organizations. This aspect underscores that companies are not value-neutral—they adhere to guiding principles that require employees to contribute in ways consistent with those values.

From a company's perspective, it is important to consider not only the economic need to let an employee go but also its responsibility toward the individual. The termination of an employee should not simply be a matter of merit or financial necessity but also of responsibility. Companies must balance justice (weighing the merit of the employee's contribution) with a duty to protect labor stability, as unemployment can cause significant personal, family, and social harm.

To achieve a balance between justice and responsibility, companies should develop policies and benefits to mitigate the negative impact of dismissal, such as severance packages, insurance, or outplacement services. This can ensure that employees are treated fairly while helping to safeguard their livelihoods.

3.3.3.2 Technological Substitution

In terms of termination, a major challenge today arises from the substitution of workers with technology. As automation increases, companies can replace human labor with machines. Several factors drive this shift, including labor regulations, production cost concerns, and the growing availability of advanced technology at affordable prices.

From a legislative standpoint, when authorities design labor regulations that favor workers too rigidly, they can inadvertently hinder a company's operations. Conversely, when companies replace workers with technology, they may contribute to unemployment—particularly if the labor market lacks a skilled workforce to fill newly created roles.

Economically, the accessibility of new technologies allows companies to weigh the cost-effectiveness of maintaining human labor against the cost of automation. The widespread availability of technology, combined with low production costs in certain countries, has accelerated opportunities for technological change. However, this shift can produce high levels of unemployment, especially when workers lack the skills to transition into roles requiring higher technological expertise.

The large-scale replacement of workers with machines can also lead to chronic unemployment due to the low employability of the existing workforce. Without the necessary skills to adapt, many workers face long-term exclusion from the labor market, potentially creating a social crisis. This slow adaptation process may take generations to resolve, delaying the return to acceptable levels of employment.

For this reason, it is crucial for companies to adopt progressive automation strategies, ensuring that employees with potential are retrained or reassigned to semi-automated roles within the same organization. This approach balances the need for innovation and competitiveness with the responsibility companies bear toward their workers and society. Automation should be implemented in a way that safeguards workers' well-being and avoids exacerbating unemployment (Redín et al. 2023).

If technological substitution is guided by the common good, it may also prevent the need for measures such as layoff bans or the controversial 'robot tax.' While these taxes aim to offset automation's impact on employment, they can ultimately burden companies, stifle business growth, and harm both employers and employees.

3.3.3.3 Fostering Inclusion and Diversity

Inclusion typically refers to public or corporate policies aimed at integrating workers who, due to certain characteristics or conditions, belong to groups that traditionally do not participate in specific industries or activities, despite having the necessary skills to meet production or service requirements.

Diversity, on the other hand, refers to the integration of individuals into the workplace who have historically been excluded due to reasons such as ethnicity, sex, sexual orientation, nationality, religion, or political views. As Domenec Melé explains, organizations can approach this issue through a taxonomy of human diversity, which categorizes it into four dimensions: ontological, individual, status, and adhesion diversities. Ontological diversity pertains to the unique personal traits of each member within the organization. Individual diversity refers to differences in age, skills, experience, and other personal characteristics. Status diversity encompasses socioeconomic and educational disparities, while adhesion diversity addresses differences in religion, culture, political beliefs, and other similar factors.

Now, when creating and implementing corporate policy, the following seven considerations and elements of judgment must be considered:

In the first place, is important to be aware of the concepts of homogeneity and prejudice. Indeed, diversity policies often arise from valid criticisms of organizations that are culturally and physically homogeneous have several problems. In particular, this refers to companies that employ workers of the same race, creed, sex, political opinion, etc. The underlying assumption is that cultural equality (the same ethnicity, sex, religion, etc.) is linked to the arbitrary and prejudicial exclusion of other groups in society. This concept applies not only to hiring practices but also to the provision of products and services to diverse sectors of the population.

In more practical terms, however, the implementation of corporate policies for diversity and Inclusion usually faces the problem of managing according to indicator. Indeed, once a company adopts a diversity and inclusion policy, it must consider the challenge to achieve better standards of diversity and inclusion according to certain goals previously adopted.

The problem with this is that diversity or inclusion does not inherently imply clear, satisfactory goals. For instance, there are always cultural characteristics that may not be represented in a group—It is, in fact, impossible to have every human condition represented in a workgroup, and not all companies can integrate workers with disabilities, as candidates may not be available on the job market.

Thus, while indicators can provide valuable statistics, setting quality standards in diversity and inclusion is challenging, when the definition of such goals may be arbitrary or impractical.

Another issue in stablishing goals is the problem of privacy. Indeed, many features that diversity and inclusion policies aim to address, such as religious beliefs or sexual orientation, are for many employees a private matter. It is ultimately up to the workers themselves whether or not to disclose these aspects.

On the other hand, corporate policies for inclusion and diversity must take into consideration the principles of merit and the way they deal with public policy and compelling legal restrictions for the company.

Inclusion and diversity policies primarily address justice—specifically, distributive and contributive justice (related to merit). In fact, this element touches on organizational justice, especially regarding diversity. Employees may perceive their own recognition or treatment as unfair, even if their merits are significant.

However, legal justice criteria (as we will explain in Unit IV) may come into play (*e.g.*, minimum wage laws) when political authorities require prioritizing certain groups in hiring practices. In the absence of labor inclusion legislation, inclusion remains a matter of maximum justice. A company that does not adopt inclusion and diversity policies cannot be deemed unjust, though it cannot be considered fully responsible either. However, this can totally change with new legislation on this matter. Political authorities can make companies incorporate such policies.

This, however, affects the necessity of firms to safeguard their competitiveness, as new and more diverse policies demand by authorities may impact on the company's performance, particularly if top technical candidates are excluded due to a lack of cultural diversity.

In such cases, the principle at stake is not only the importance of being competitive, but also, and more important, the limits of corporate autonomy. In our opinion, for instance, even when legislation mandates inclusion, it cannot supersede a company's right to determine its hiring practices. Corporate autonomy, grounded in private property rights, allows companies to manage productive resources as they see fit. However, companies must also ensure that workers have the essential technical skills required for the role. This can sometimes hinder progress in diversity initiatives due to equity concerns, not just for company owners but also for other employees who depend on the company's long-term success.

However, corporate autonomy has certain limits. In fact, form the perspective of the common good, the problem of inclusion and diversity are ethical principles derived from the importance of active participation in the workforce. Excluding people from work, whether for arbitrary or circumstantial reasons, prevents individuals from developing in an essential aspect of life—work itself. This not only negatively impacts the excluded individuals but also society, which loses the opportunity for their development to contribute to the common good. Accordingly, inclusion and diversity policies, grounded in maximum justice and subsidiarity, can provide opportunities for personal growth and contribute to the common good, particularly for those who, despite being capable of contributing, have been excluded for arbitrary or circumstantial reasons.

In contrast, corporations should avoid what can be described as instrumental inclusion and diversity—that is, efforts to improve diversity standards justified not as ends in themselves, but as means to achieve economic or reputational benefits. For example, the inclusion of women in corporate leadership is often valued not only for introducing different leadership styles but also for bringing the broad perspective necessary for effective policymaking. Likewise, integrating people with disabilities is sometimes pursued primarily as a strategic move, particularly for branding or corporate communication purposes. While this approach is not inherently unethical, it remains incomplete if it fails to recognize inclusion and diversity as values in their own right—rooted in the intrinsic worth of personal work.

Finally, it is important to consider that once firms have achieved significant diversity and inclusion, a basic principle is to act with care and respect. For instance, diversity and inclusion policies grounded in care should not only involve integrating workers with disabilities but also ensuring the physical accommodation necessary for them to perform effectively. In this sense, respect and care extend to creating organizational environments that foster a corporate culture of cooperation and friendship among individuals with different backgrounds and perspectives. Such culture, in fact, should also integrate the guidance needed to safeguard several principles such as freedom of expression.

The experience shows, so far, that companies can either prohibit the expression of such characteristics (as in some European companies that ban religious symbols) or, in totally different manner, they have created an environment of respect, where such expressions are allowed. Respect, in this sense, has been defined under the importance of defending the right to disagree. This does not conflict with care when the expression of differing opinions is not inherently offensive but merely discrepant. For example, it is not inappropriate to argue that Nazism is wrong, but it would be disrespectful to label a neo-Nazi employee a murderer.

Manifestation: The principle of freedom of expression also provides criteria for limiting policies of care and respect that are ultimately unjust. It is unjust to force individuals to express or physically manifest principles they do not share. For instance, a Middle Eastern company should not force employees to fast during Ramadan if they do not wish to.

Position: Lastly, it is advisable to focus on one's own position rather than imposing one's views on others when discussing values or opinions unrelated to the organization's functioning. For example, in a conversation about gender identity, it is preferable to state one's personal beliefs, such as being a practicing Jew, rather than making judgments about others' beliefs or positions.

3.3.3.4 Enabling Whistleblowing or Speak-up Policies

The concept of whistleblowing was introduced in business ethics to describe the practice of professionals raising concerns about weaknesses or defects in product design that could eventually harm consumers (Ceva and Bocchiola 2020). Over time, the term expanded into the so called Speak Up practices much more oriented to include the disclosure by company members of corruption, fraud, or abuse within the organization, all aimed at promoting changes and improvements in corporate practices.

However, contemporary Speak Up safeguards the importance for whistleblowers to take certain consideration before any accusation. Accordingly, it is crucial that a whistleblower's accusation be framed within a rational assessment of loyalty. This means the whistleblower must be confident that disclosing the information is intended to uphold or strengthen the company's mission, values, and principles—especially when these are explicitly stated. Only if the whistleblower believes that the reported acts violate these corporate principles is their action justified and considered loyal. Another basis for whistleblowing is complicity, where the action is justified not by loyalty to the company but by the right to protect oneself from involvement in unlawful activities.

As for the company, Speak Up policies incorporate all the needed resources to prevent any direct or indirect retaliation against the whistleblower, underscoring the importance of anonymity. Non-confidential information related to conflict resolution does not imply public disclosure but should be handled in specific corporate forums where the source or individual providing the information is known.

In this vein, to protect the whistleblower's anonymity, a confidential conciliation process must be initiated. This requires involving a third party, that is, the so-called ombudsman—a term derived from the Swedish *umboths* (commission) and *mathr* (man), first used in 1809 to designate an official in the Scandinavian political system tasked with public advocacy. In the United States, the ombudsman role was introduced in companies in the 1960s to handle consumer complaints and reduce litigation costs. Within corporations, the ombudsman primarily focuses on conflict resolution and typically has direct access to company management and the board of directors, ensuring effectiveness in resolving disputes (Robbins and Deane 1986). The ombudsman holds responsibility but has no formal authority and is usually empowered with investigative and auditing capabilities to better understand the nature of complaints received.

3.4 The Ethics of Leading People at Work

The role of leaders in organizations must also be framed according to the principles of autonomy and self-determination. From this perspective, leading people is, in a neo-Aristotelian sense, a political matter among free individuals. As Aquinas states (*S.Th.*, II-II, q.50, a.2), those who are led by others should follow instructions in such a way that they are guided by their own free will. However, for leadership to be ethically sound, it must not only safeguard the autonomy of those who are led but also ensure the good of the task at hand. This implies that—from the leader's perspective—what is asked must promote both the personal good of the employee and the common good of the organization. The common good, in this vein, does not merely align with the personal good—it is, in fact, the personal good of each employee, and vice versa (Llano 1982).

Additionally, human actions are good as they seek human goods by the correct means of command (Murcio and Scalzo 2021). Hence, it is important for leaders to motivate employees to follow instructions by using the appropriate methods that can safeguard both autonomy and the personal and common goods at stake. This is, in Aristotelian terms, the ethics of persuasion. For Aristotle, "the use of persuasive speech is to lead to decisions ...This is so even if one is addressing a single person and urging him to do or not to do something ... "(Aristotle, *Reth.*, 1391b-1392a) in matters of good and utility according to which the means of persuasion can be morally assessed (Aristotle, *Reth.*, 1363b). In this vein, continuous Aristotle, we might say that

> There are, then, these three means of effecting persuasion. The man who is to be in command of them must, it is clear, be able (1) to reason logically, (2) to understand human character and goodness in their various forms, and (iii) to understand the emotions-that is, to name them and describe them, to know their causes and the way in which they are excited. It thus appears that rhetoric is an offshoot of dialectic and also of ethical studies. (Aristotle, *Reth.*, 1356a-1356b).

In this sense, persuasion has a higher moral ground than simply forcing or deceiving people to make them do something. However, this implies some moral commitment from a leader, that is, respecting followers' autonomy and their personal, common good. According to Aristotle,

> ... rhetorical persuasion is affected not only by demonstrative but by ethical argument; it helps a speaker to convince us, if we believe that he has certain qualities himself, namely, goodness, or goodwill towards us, or both together. Similarly, we should know the moral qualities characteristic of each form of government, for the special moral character of each is bound to provide us with our most effective means of persuasion in dealing with it. We shall learn the qualities of governments in the same way as we learn the qualities of individuals, since they are revealed in their deliberate acts of choice; and these are determined by the end that inspires them. (Aristotle, *Reth.*, 1366a)

In fact, when leaders are granted by followers with a recognition of their goodwill—demonstrative and ethical arguments in the words of Aristotle—this can be called authority. In contrast, leaders possess only power when they direct others according to their own instructions, without the participation of the autonomy of their followers or motivating them with goals that are not genuinely personal or common goods.

3.4.1 Power and Authority in Leadership

Power (from the Latin *potestas*) is rooted in external motivations, meaning it compels individuals to act not because they personally aim to achieve a specific goal, but due to the consequences of failing to do so. A manager's power establishes a relationship of obedience, where employees feel obligated to follow orders, even if they do not understand or agree with the underlying reasons.

The concept of power, however, does not arise in the context of management sciences, but rather in political philosophy. Power, in the political sphere, relates to the legitimate and legal use of force (e.g., the state's legal monopoly on violence). This refers to the coercive measures by which the state and political authority can compel in accordance with the law. This argument, which has its roots in Roman law, reveals how authorities can command citizens in order to preserve social peace and the public good, even when some citizens are not really committed to said ends.

But, unlike political power, power in corporate management is based on a relationship between capital and labor (many times abbreviated as 'K+L') since whoever owns capital also owns the means of production. In this context, workers who contribute their labor or work can be incorporated or removed from the use of the means of production, that is, they can be hired or dismissed when those means are legitimately and naturally privately owned. Hence, an owner's resources, or those to whom he/she delegates power (managers, administrators, or directors) can coerce workers with the threat of firing. According to Arendt (1958) necessities of life can coerce men and hold them in their power.

However, managing based on power is not necessarily illegitimate from an ethical point of view; indeed, companies require power for decision-making and to safeguard the company's best interest. The problem arises when power is exercised as the only management tool; this approach is incomplete and, hence, qualitatively inferior.

At the same time, understanding management as power is the basis of what is known as transactional leadership, which is, in turn, based on economic and contractual criteria that are upheld by the right to private property and by employees' effective contribution to the company. Thus, power-based management promotes transactional leadership styles, as it only moves by virtue of a transaction that is based on a simple process of trading resources. This, again, is not bad in itself, but it is partial because it does not necessarily imply a form of authority and does not ensure the exercise of freedom at work.

Unlike power, authority (from the Latin *auctoritas*) is based on the ability to guide others through a shared deliberative process, where everyone embraces the common good as their own (Ciardi and Scalzo 2022). In such frameworks, while the initiative may rest with the manager, employees personally adopt the reasons for action and the objectives at hand. In practical terms, authority-based management motivates employees through recognition and trust in the manager's rationale, prompting them to act out of genuine desire rather than mere obligation. This approach fosters a form of managerial responsibility rooted in authority. In other words, effective people management should aim to enhance individuals' deliberative capacity—improving their ability to make decisions and their willingness to act, thereby strengthening managerial authority.

Authority, in line with Aristotle ideas on demonstrative and ethical argument, can be distinguished in its technical and moral dimensions. The former is granted to leaders when they are trusted in their know-how, their capacity to solve technical problems. In fact, this authority can simply consist in recognition of acquired knowledge or of having witnessed how the manager has solved technically complex issues over time.

However, this cannot be reduced to a purely technical matter, especially because it involves the element of authority. In this regard, technical authority is granted within the context of a learning relationship between leaders and followers; without this, it loses its connection to autonomy. Those who are recognized with technical authority are characterized by continually training and enhancing their subordinates' ability to solve technical problems in the same way they do. Therefore, having technical expertise alone is not sufficient to be considered an authority in these matters; it also involves sharing this knowledge and making it accessible to others so they can internalize it.

Moral authority, on the other hand, has to do with trusting a leader or manager's intentions. It also involves the managers' experience and biography, i.e., aspects of a person's life that provide certain guarantees of good intentions and of genuine interest in employees' personal growth beyond technical-productive aspects. Hence, moral authority is that which allows for relationships of trust and is achieved on the basis of transformational leadership. It includes a recognized willingness to improve those under management, as well as to transform them for the better, and this applies both to professional/productive matters *and* to personal/company-related ones associated with employees' lifestyles (considerations like family, rest, savings, specialization, career, among others, come into play).

3.4.2 Leading for Fellowship

Ethical leadership in terms of technical and moral authority is featured by achieving higher standards of integration (engagement), meaning (meaningful work), and empowerment in the company's members. In other words, these three conditions can be considered as indicators, or an important outcome of a direction based on technical and moral authority (Al Halbusi et al. 2025).

Engagement or belonging in the workplace refers to the experience of authorship, i.e., effective participation in the processes and achievements related to one's own activity. As we have explained before, this is not only a matter of merit in terms of results, but also of the personal initiatives that require obtaining the necessary resources to achieve given results. Thus, participation should translate into an awareness of shared responsibility for the company's common good and in the satisfaction of having participated in the organization's achievements. Thus, engagement does not just involve communication policy or certain cultural changes; it also includes the verification of achievements on the part of a company's members.

On the other hand, we can maintain that work is meaningful to the extent that it is desired and, moreover, because I identify with that which I desire. Highly meaningful work is usually associated with carrying out a vocation. But because the vast majority of people do not manage to match their work up with their vocational interests, or simply do not know what their vocation is, meaning is usually acquired through general meaningful goods, such as service, efficiency, knowledge applied to work, etc. All of this, however, involves considerable levels of discernment, deliberation, and the acquisition of meaningful goods. This implies that workers not only value their work individually, but as something that is done as a whole, with a sense of collegiality and collaboration.

Empowerment involves not just autonomy, but also the capacity for initiative and delegation in decision-making. This implies the provision of organizational resources for decision-making and bringing necessary technical knowledge to the table to decide what one's tasks will be to achieve a given end. Thus, empowerment is essentially a process that increases workers' responsibility when it comes to improving operational, organizational, commercial, etc. results.

Yet, the relationship between authority and the achievement of these three conditions of good work is established under the logic of purpose within the company. In other words, both the connection, the sense of work, and empowerment are possible under the umbrella of an organizational purpose that can be communicated by the leader and integrated by those who are directed. Thus, the content of the purpose, when well conveyed, is what allows the members of the company to feel connected to the project, give meaning to their work, and understand the significance of spaces of autonomy.

Following Aristotle, purpose is a *telos* (τέλος) or the various aims of what is pursued through practice. An aim is 'that for which' something is done (*see Physics*, II,2) by a person or, extensively, by a group of people. The Stagirite explains that a person can consider 'health is the reason for my daily walks'. Thus, both what is sought ('being healthy') and what we have set as means to the end ('the walks'), have a teleological condition that allows them to constitute a purpose (Aristotle, *Physics*, II,3). Purpose, in this sense, is a coherent conjunction of aims in a specific activity.

For Aristotle, everything we do, we do with a view towards an end (a good), and not just one for each action, but often several ends in the same activity (Aristotle, *NE*, 1097a). Thus, a single purpose can be composed of diverse and interconnected ends. However, even if they are multiple, they can be classified into two types: ends in the order of effectiveness (instrumental goods) and noble ends (goods in themselves or, failing that, apparent, meaning mistakenly considered as truly good) (Aristotle, *Physics*, II,3).

Purpose in the order of effectiveness refers to the productive processes and their causal relationships that, according to Aristotle, occur, for example, in the construction of walls for a roof where the heavier elements must compose the wall, otherwise the structure collapses (Aristotle, *Physics*, II,9). These productive purposes have, in turn, two constituent elements, namely, the functionality of the processes (effectiveness) and the assurance over time of the resources that allow production (efficiency); elements that constitute the base of technical authority,

For Aristotle, the instrumental goods that ensure the efficiency dimension of purpose are money, honor, and power. These, in turn, are the resources that are secured within the institutional dimension of the company (MacIntyre 2016). However, it is important to note that these instrumental goods are proper goods, because they are a condition for a good life and their acquisition is essential for overcoming material or other impediments that hinder a fulfilling life.

On the other hand, all ends in the productive order are achieved for something other than themselves, like exercise for health, or a house for shelter, meaning they are always intermediate ends. But nobody can live solely in the function of means and instrumental goods, that is, seeking ends for further goods so that same goods subsequently become means for other goods. This, we can say, is a condition of despair, akin to that of Sisyphus. Therefore, it is crucial that there is a good in itself, something noble, that is achieved during the action (not at the end of the action) and is sufficient in itself (*see* Aristotle, *NE*, 1119b-1120b).

The only thing that can be pursued for its own sake is the good of another person and, by extension, the common good (Aristotle, *EN*, 1156b). These goods are essential to what Aristotle refers to as life's purposes (*NE*, 1160a), which are achieved when one lives in accordance with virtue and without obstacles to personal and common development. More specifically, we must say that noble goods are characteristic of friendship, that is, seeking the good of the other or others, but for the sake of the other or others, without an instrumental character, in other words, benevolently. For the company, this noble good is, indeed, friendship at work (Llano 2000).

According to the concept of purpose as an Aristotelian *telos*, a leader becomes an authority when they achieve, both in their relationship with those they lead and in the interactions among them, the realization of effective goals (that is, in the logic of technical authority) in harmony with noble ends or friendship at work (that is, in the logic of moral authority). In other words, an Aristotelian approach to moral authority in organizations can be understood through the logic of friendship at work.

According to Llano (2000), friendship in the company is characterized by a unity of three dimensions, namely, companionship, sympathy, and a community of interests. The former, companionship, refers to the co-incidence in the same activity, meaning mutual causality or involvement in a common task. As we have seen before, collaborative relationships are fundamental for the existence of the company. However, not every form of cooperation is companionship. In no case is companionship an arrangement of individual tasks - as in the productive logic of Fordism—but rather participation together in work teams. Companionship, therefore, is not a concatenation of individual processes, but a joint activity in which the final result of the task is the outcome of diverse contributions from all members of the work group.

Sympathy, on the other hand, is a characteristic of qualitatively superior and usually more effective companionship relationships. From the Greek *sym* (*σῦμ*) and *pathos* (*πάθος*), sympathy means 'to feel or suffer with another'. It is not simply 'getting along or liking each other', like compatibility of personalities. For Aristotle, in fact, *pathos* does not refer only to a feeling, but to the effects or feelings that affect judgment (Aristotle, *Rhet*, I,2). In this way, companionship is strengthened by becoming a 'community of tastes in decisions and actions', that is, a common feeling that is incorporated into a common judgment. Hence, companions are characterized by deciding together, and better yet, when the feeling regarding those judgments is equally shared.

Finally, friendship at work is also a community of interests. However, this community does not mean that friendship in the company requires two or more people working together to have interests in the same things, or to be interested in various things in the same way. The community of interests—as explained by Llano—lies more in complementarity, that is, arousing interests that would remain unfulfilled without mutual assistance. Hence, the formation of teams requires a certain level of diversity, not like a melting pot, but in the manner of a mosaic (Llano 2000).

In short, moral authority in Aristotelian terms is possible as it is based on the establishment of friendship relationships, which in turn constitute the noble end of the purpose of any work organization. Hence, only the logic of friendship (comradeship, sympathy, and common interests) can achieve jobs characterized by their engagement, meaning, and empowerment; and only in this logic is it possible for those being led by managers with authority, not just power.

Finaly, is important to consider what we can call false authority. As said before, authority is achieved when it is granted by those being led. However, authority is genuine not only because of the recognition from those being led, but also because such recognition is founded on correct moral principles. In fact, a recognition of authority can be false, that is, based on deceit—as in charismatic leadership—or the lack of moral character of both leaders and followers—as, for instance, in the practices such as paternalism or competitiveness.

Deceived leadership, indeed, often derives from strategies used to build trust, hiding individualistic purposes. This, in fact, is a matter of manipulation. From the Latin '*charisma*', and from ancient Greek *χάρισμα*, from *χαρίζεσθαι*, this concept means to please or do favors but associated with the ability of some individuals to attract or fascinate. Manipulation, in this sense, requires charismatic skills that allow for the creation of false trust without the genuine interest of others.

In a different manner, paternalism is a false form of ethical leadership not as it deceives, but as it removes autonomy for what is considered the benefit of the follower. Paternalism, in this sense, guides well, conceives, and seeks a good end, but it does not allow for spaces of development, real autonomy, and, ultimately, the exercise of practical virtues at work.

Competitiveness, on the other hand, occurs when both leaders and followers lack virtues and a sense of excellence. A false concession of authority to a manager occurs in the case where instrumental ends (money, power, prestige) are promoted and valued as if they were ultimate ends. As MacIntyre (1981) explains, relationships around these instrumental goods are always established in a logic of competition, not so much collaboration. In this sense, in firms where the authority-follower relationship is based on instrumental goods, relationships are often shaped by the logic of competition—particularly commercial competition. This dynamic integrates money as revenue, power through market dominant positions, and prestige in terms of reputation. However, commercial competition or competitiveness in itself is not inherently bad. The mistake lies in considering that all activity in the organization is reduced to it. Thus, what we could call 'competitivism' is simply a reduction of all activities in the company to competition for the attainment of economic benefits, prestige, and power which, in the case of management, can lead to a false form of leadership.

In short, false authority (or an authority granted erroneously) places instrumental goods as if they were goods in themselves, turning relationships in the company into competition, rather than friendship.

3.5 Fairness in Compensation

Compensation is clearly a matter of ethics and, particularly, of justice. This is, in fact, something quite clear in those discussions in which the fairness of certain forms of salaries are controversial, such as wage gaps, poverty and minimum wages, payment discrimination, or even Gini´s distribution of wealth, among other issues associated with compensation (Pinto and Scalzo 2024). All of them, in fact, are a matter of being fair, when justice is precisely the good habit—virtue—according to which one gives to another what he is entitled to receive (Aquinas, *S.Th.*, II-II, q.58, a.1), allowing the possibility of all community members to be active in establishing better social, economic, and political relationships.

However, an Aristotelian approach to compensation fairness is not merely about applying a general definition of justice without nuance. While all human relationships can be evaluated through the lens of general justice, different types of relationships, with their unique characteristics, require specific criteria of justice to better understand in what sense particular compensation schemes are just. Therefore, the first step in assessing compensation fairness is to establish a distinction between the types of compensation and the specific form of justice that applies to each.

3.5.1 Three Parts of Justice, Three Kinds of Compensations

Compensation or remuneration is the exchange of checks, bills, or other securities between individuals and institutions—typically involving the periodic settlement of reciprocal credits and debits—related with the performance of tasks, services, production outcomes, or any other specific aspect of human works. In this context, compensation is essentially a matter of debt owed for performed or to-be-performed work. Therefore, in all working relationships there are creditors and obligations, all of which, in turn, pertain to justice when reciprocity and equivalence between payment and service are essential features of said relationships.

However, as compensation occurs in different contexts of work, assessing the fairness or justice of them cannot be considered in general terms, but in accordance with the kind of relationship in which different agreements of work take place. In fact, such distinctions are shown in the way we refer to several forms of compensation: those compensations proper to contexts of cooperation—such as the firm—are referred to as wages or salaries, while that which take place in a commutative or exchange context—i.e., the market—are called honorary fees or just fees.

The latter were originally a form of compensation called as such because they honored the work of someone who has rendered a service—formerly they were referred to as 'honorary salaries,' perks paid to doctors, lawyers, or other liberal professions. Thus, honorary fees related to a form of relationship that took place in markets and, therefore, established under the logic of mutual benefit proper to commutative justice. For Aristotle and Thomas Aquinas, this form of justice is that which takes place in the tradeoffs between two people (*see* Aristotle, *EN*, 1131b-1132b; Aquinas, *S.Th.*, II-II, q.61, a.1) or between individuals or groups (Finnis, 2011a). Commutative justice is present in exchanges of products and services that pertain to trade, usufruct, loans, deposits, or rents (Aristotle, *EN*, 1131a), that is, in any activity in which exchange takes place (Aquinas, *S.Th.*, II-II, q.61, a.2).

Now, justice in the market is established on the basis of two principles: Mutual benefit and proportional trade-offs. The former is indicative of the benefit received by the parties involved in both benefits psychologically and objectively from the exchange; they feel their need is satisfied and that satisfaction is objectively good. Proportional trade-offs refer to the need for goods and services to have a proportional value related to the needs of each party. This proportionality is made objective with money: An exchange value represents a criterion of proportionality when the price paid is what one is willing to pay for the satisfaction of a need.

On the other hand, wages and salaries have their etymological origin in the Latin *solidum* and *salarium*, respectively. The solidus was a Roman coin equivalent to 25 gold denarii. *Salarium*, on the other hand, indicated 'payment in salt,' which was a highly prized resource in ancient times because it preserved food. Both forms today meant the same, that is, compensation owed for performing in a certain position—many times a job performed according to a contract that sets out time frames (hours or days) and/or results (piecework). A salary, in this sense, pertains to an activity performed as a whole—similar to the work of a Roman army. In other words, it is part of a broader cooperation framework in which the efforts of all members of the organization converge. Thus, compensation in the form of salaries follows the criterion of merit based on the performance of the group, aligning with the principles of contributive-distributive justice.

As said before, distributive justice (Aristotle, *NE*, 1131a) concerns only those relationships between an individual and the community in which the same individual has been participating in an active manner. The criterion of justice is, in this case, merit (Aristotle, *NE* 1131a; 1131b; Aquinas, *S.Th.*, II-II, q.61). Accordingly, distributive-contributive justice gives to each member of the firm what is due, that is, in a proportional manner considering the contribution of each single member to the good of the community. (Aquinas, *S.Th.*, II-II, q.61, a.1-a.2).

In this sense, when workers join a company, they do not establish a commercial relationship with their employers but rather enter into a relationship based on cooperation—as we will explain, the only connection with the market is associated with the amount of money in compensation. Employees, in fact, do not sell the results of their work to the company monthly or at any other regular interval. Instead, they cooperate in collective activities with other members of the firm. For this reason, they are entitled to a proportional reward, which can be claimed from the company's overall profit according to merit.

Hence, it is correct to sustain that firms are developed under contributive-distributive relationships. Neither individual employees nor companies can achieve results without a cooperative effort, whether because of the availability of economic and technical resources provided by the organization, or because of the productive-management capacity provided by employees.

However, cooperation does not mean that each effort has the same value. Distributive justice, as based on merit, does not imply egalitarianism in the distribution of benefits. In other words, compensation according to distributive-contributive justices implies a distribution of the company's profits in terms of each worker's merit. Equitable differentiation is necessary since distribution can never be equal if it is to be just. Justice in distribution, therefore, must take into account each member's preponderance, importance, and effectiveness in achieving the firm's objectives. This means that merit refers not just to the effort, knowledge, experience, and responsibility put into a task, but also to the results obtained and its relative relevance. In this sense, distributive justice is associated with both contribution and results, and according to this, it is fair that some company members earn more than others.

In a different manner from salaries and honorary fees, compensation can be also defined as income, which concerns the criterion of legal or political justice, that is, the common good of a society. As said before, legal-political justice establishes the criteria of fairness applied to the relationship between citizens and private corporations with the State (Finnis, 2011a), whose role is associated with the good of society. Legal-political justice, in this sense, prioritizes those goods that are more important for society, by means of safeguarding essential goods and values that are fundamental to culture and society.

Accordingly, income refers to a societal-level concern. For this reason, in terms of justice, the parties that receive incomes are not considered as individual workers, but citizens or households that, from a social perspective, need, for instance, a minimum quantity of economic resources to sustain themselves in order to allow social peace and a dignified living. Naturally, this includes income associated with salaries and fees—as in the case of minimum wage policies implemented for reasons of public good—but also other monetary and non-monetary benefits provided by the State or other non-profit associations. In this sense, income takes into account public education, unemployment insurance, and family support policies, among other benefits funded by other members of society through mandatory taxes.

3.5.2 Justice and Responsibility in Compensation

As mentioned, honorary fees, salaries, and income must be evaluated according to the different parts of justice: commutative, contributive-distributive, and legal or political justice, respectively. In other words, honorary fees should be analyzed based on commutative justice according to the principle of mutual benefit and proportionality. Salaries, on the other hand, should be assessed in terms of distributive-contributive justice under the criterion of merit. Income, for its part, must be ethically understood in the context of legal-political justice, which addresses broader societal and legal under the logic of public good of society.

However, the fairness of honorary fees, salaries, and incomes must be assessed not only in accordance with said principles of justice, but also in considering the economic amount of each compensation (monetary and/or non-monetary); otherwise, the normative analysis of compensation would be incomplete. This idea, in fact, responds to the fact that, on certain occasions, paying reduced amounts of money can be unfair.

What makes an amount of money fair or unfair in terms of compensation?

Based on any criteria of justice, it is impossible to establish with precision a fair amount of money in compensation in general terms. Among other reasons, this impossibility derives from the fact that each criterion of justice is a permanent framework for analysis. But, on the other hand, the value of money and the costs associated with a good living change over time, and they are different in different economies and cultures. This is not to say that an amount of compensation, whatever it may be, is unrelated to justice. This means that its fairness cannot be derived from a criterion of justice, but it has its own ambit of analysis that can be added in a full assessment of justice in compensation (Moriarty 2012, 2014).

In our view, even if such considerations are not explicitly found in Aristotle's writings, they can be traced to Aquinas' arguments on fair pricing. Accordingly, we can sustain in the first place that a concept of a fair market price includes the cost of production and labor, factors that naturally vary across places and times (Finnis 1998). Therefore, fair economic value must be often understood according to what a society deems reasonable to pay for goods and services according to what that same society has considered a good living standard. In this context, a just person would assess the fairness of a price by referencing market customs—what nowadays is reflected in labor market value for different jobs—that is what the majority of people typically pay for a given product or service (Aquinas, *S.Th.*, II-II, q. 71, a.4).

Thus, the economic value (or price) reflects the broad spectrum of values society assigns to its preferences and wants, which are considered reasonably fair not because they are concurred, but because they align with what society recognizes as a humane and dignified way of living in its minimum standards. According to this, referential market prices are not inherently determinative of justice. Instead, they act as 'proxies' or references of fairness that each person should apply in economic transactions—criteria typically based on what individuals believe is necessary to secure a good and minimum standard of living. In other words, personal judgments of a particular society reflected in market prices form a spectrum for fair pricing, naturally allowing for some margin of error. Thus, all economic transactions—including wages and fees—are mostly fair when they cover a worker's costs of living or, in other words, that which is required to perform a job and live reasonably well. In this way, the economic value of work depends on ethical terms of what we understand for a good living in its basic standards. The economic cost of living, however, is not assumed in particular to each employee, but in reference to what is usually paid. In fact, employers are not morally obliged to cover the costs for a worker to live well, but they are responsible to pay what is usually paid for that job—a general parameter—as this amount covers essential needs, especially the case of low-income workers.

In short, the market itself determines the economic value of just payment (including fees and salaries) as a reference (benchmarking). Paying what is usually paid tends to be just as this assumes that market players have established what is in their best interests. This idea, however, does not imply that what is established by markets is normative; it is only a reference. Indeed, poverty wages, which are unjust from a legal-political perspective, may be in line with the market standard. Therefore, it is correct to sustain that market, even though essential moral assessments of the economic value of compensation are limited.

3.5.3 Practices Associated with Compensation Ethics

Practices in matters of compensation can also be classified as wage agreements—that is, contractual practices, arrangements, or any other form of working relations that are more or less acceptable according to the principle of justice in compensation. The agreements we will review here are among the most common, though there may be others we do not address directly, such as *ad honorem* work relationships (internships, religious services, volunteering, etc.), basic universal salary, public service compensation, and so forth.

3.5.3.1 Paying Minimum Wages

Minimum wage involves three important considerations, as follows: The related circumstances (can a worker improve his situation by earning a minimum wage?), the duration (is it acceptable for a person to earn minimum wage for a prolonged period?) and the company's profitability (is business activity that is profitable over time justified when its labor costs are associated with a minimum wage?).

To address these considerations, we must first establish that a minimum wage is the result of a public policy that mandates the payment of a basic hourly or monthly amount. These wages are, in terms of amount, a matter of legal justice, as the calculation thereof looks to parameters associated with subsistence or good living (hence the use of the market basket or other more or less objective indicators).

Now, one issue with minimum wages is the fact that, in many developing countries, the problem is not so much the amount, but its permanence over time. Because of minimum wage legislation, many people in poorer economies may improve their situation because they achieve more regular income, social benefits, and they may even get out of poverty. However, this does not justify the maintenance of a minimum wage; in other words, a worker earning the legal minimum over a long period of time. It is partly the company's responsibility to increase overtime the income of a worker who is hired at the minimum wage, a responsibility that has to do with the dignity of work and that must be taken into account when the company can pay more, and the worker can to some extent increase his contribution.

Thus, companies, in collaboration with workers, also assume a sense of responsibility that materializes through their participation in training processes, initiatives to learn new tasks, etc.

At the same time, a business model that incorporates a large volume of workers at the minimum wage over a prolonged period as a strategy for reducing costs and increasing competitiveness cannot be justified. Thus, the company's profitability cannot depend on cost controls that keep many employees working for minimum wage on a permanent basis, especially when the minimum wage is also poverty wage.

In addition, these business models often do not allow workers to build a career, as there is a skills gap between the operational and managerial segments that prevents career development. Most operative workers do not have the opportunity to earn more than the minimum because the way the company structures work does not allow it. This can be seen in fast food and retail chains, among other industries.

3.5.3.2 Paying Poverty Wages

These are wages or salaries that, even though they are equal to or above the minimum wage, do not allow workers and their families to live a decent life. Poverty wages can be the same as the minimum wage (as in countries such as Chile), but the minimum wage can also be considerably higher than a poverty income (as in the case of Spain). The minimum wage and the poverty line do not necessarily coincide. Poverty wages are often associated with jobs that have almost negative profitability. In other words, workers contribute less than they earn.

Now, poverty wages should be corrected for a variety of reasons, but there are two especially important reasons to consider initially: Productivity and dignity.

In terms of productivity, poverty wages, as mentioned, can signify a negative productivity problem.

In this sense, an increase is not justified. Faced with this scenario, companies must subsidize, meaning they must provide resources and policies to improve the contributive capacity of workers. This is a principle of integral responsibility, as it seeks the development of workers as workers, citizens, parents, etc., and should be applied urgently for the neediest.

However, there are certain limits to organizational subsidy policy that depend on a company's capacity to develop (since no one is obliged to do the impossible) and that of workers. Not all workers are capable of improving their contribution to a company.

In terms of dignity, poverty wages are unjust when a worker with consistent work activity cannot rise above his present condition.

This means that work becomes a mere activity for survival, and this is contrary to the very nature of work as an essential human activity. However, we should not conclude that workers in such conditions are immediately prepared to earn more.

Thus, dignity demands concern on the part of employers and authorities, but this does not mean that wages must be raised at any cost. Indeed, the issue is that, in consideration of human dignity itself and the essential value of work, workers must earn wage increases, because simply giving away benefits as charity is also contrary to dignity as

3.5.3.3 Offering Variable Wages

Variable wages are a labor agreement that has both positive and negative aspects. On the positive side, a variable salary responds to a criterion of distributive justice. In other words, compensation depends on the worker's effort.

This, however, forces us to consider two potentially negative aspects of this type of compensation: Companies do not always offer the necessary conditions for workers to achieve the required goals and, consequently, an adequate income. If this occurs, a true relationship of distributive justice fails to exist.

The second refers to the way in which a company can safeguard against poor performance that occurs for reasons beyond its own efforts and those of employees. For example, when there are economic or social crises, health problems, etc.

Companies, as associations and working communities, play a role in labor stability from which follows, among other aspects, the need to cooperate not only on production but also in exceptional situations. In this regard, variable compensation against results should be implemented with a lot of solidarity that translates into base salaries or other forms of stability and protection.

In the history of companies, this solidarity principle has been achieved through workers' associations (savings cooperatives, for example) or through union activity which, together with companies, protects workers in difficult times.

3.5.3.4 Outsourcing

Outsourcing is not the same as subcontracting. The latter refers to externalizing a regular or sporadic company activity. Outsourcing, on the other hand, refers to the practice of incorporating the service of another company that provides work that the contracting company already performs on a regular basis with its own employees. Thus, subcontracting companies can not only grow their operation by producing savings regarding the costs associated with new workers but can also free themselves from some of the labor obligations to which they are subject by law, because the subcontracted workers are not considered to be under their responsibility.

This practice—often sanctioned when the outsourced company performs the same activities as the requesting company's workers—usually produces problems regarding justice and working conditions.

In terms of justice, when two workers do the same thing but earn differently and/or do not have the same benefits, there is clearly an equity problem.

At the same time, outsourced workers are considered on the basis of commutative justice criteria, and this is a distortion of the employment relationship.

All in all, outsourcing practices are usually considered unjust. However, when they are necessary (as in the case of sporadic staffing needs or subcontracting), the working conditions, as well as the compensation of outsourced workers, must be safeguarded.

3.5.3.5 Paying Ethical or Living Wages

Ethical wages refer to both public policy and social movement. The former usually seeks to incentivize and reward companies that pay higher salaries to workers who usually receive low compensation. Thus, instead of implementing changes in minimum wage law, it is left to the discretion of the company itself and its ability to pay.

The social movement aspect of ethical wages usually involves campaigns associated with non-profit institutions that promote increasing low wages in the country or as has also happened, encourage changes in company policies in order to reduce internal wage gaps through redistribution processes.

Now, as we have said, a salary is ethical not just because of its amount (since we cannot deduce a fair amount from a criterion of justice), but rather regarding the minimum standard of living found in society. Thus, ethical wages seek to allow people who work regularly and are productive to live in good conditions, and for this to be assumed not only by the State, but also by companies themselves. Now, if we were to accept the idea of an ethical wage, it should be established on the basis of justice criteria, but it cannot be a specific indicator of the amount to be paid, as this is rather arbitrary.

3.5.3.6 Salary Gaps

This refers to very high salaries, often in the millions of dollars, that produce a particularly wide salary gap within a company. This is the case of many American corporations where the CEO has a high income and compensation while those workers who earn the least receive minimum or very low salaries. Now, these cases do not imply a problem associated with the amount that a manager receives, as this may simply be associated with the market and such compensation may not affect the continuity of the company, nor its profitability. Nor is it necessarily a gap problem, unless it has a negative impact on the perception of organizational justice and, thus, on the company's stability. But if that does not occur, the gap is not truly an organizational problem.

In reality, such wages are not in themselves unfair when the company fulfills all of its institutional responsibilities. Such salaries are, on the contrary, unfair when those who earn the least within the same company are not able to live with dignity and this can be corrected by means of income redistribution in such a way that the higher salaries do not cease to be competitive and increasing wages for lower salary earners is done in a way that corresponds to their contributions to the company.

In other words, the argument is established on the basis of low wages, which, in addition, could be improved under certain conditions, making those who earn more earn less, without ceasing to receive competitive compensation. The problem with wage gaps is an issue that arises from identifying very low incomes that have real potential for improvement, not from compensation policy for managers.

3.6 Management and Agency Ethics

An Aristotelian theory of management is grounded in a distinction between economy (οἰκονομία) and chrematistics (χρῆμἄ) (Aristotle, *Pol.*, 12561a-1256b). The former can be understood in modern terms as administration, management, or even agency. According to Aristotle, economy, or household management, is the art of acquiring property and using it to secure a good life. Similarly, Aristotle views work management as the provision of necessary resources for workers to fulfill their tasks (Aristotle, *Pol.*, 1253a).

On the other hand, chrematistics refers to the art of wealth acquisition—a concept akin to modern finance—and, as Aristotle asserts, it forms part of management (Aristotle, *Pol.*, 1256a-1256b). However, chrematistics suggests a sense of limitation, perhaps because excessive wealth can undermine a good life or, as Aristotle explains, "the amount of property which is needed for a good life is not unlimited..." (Aristotle, *Pol.*, 1256b). Wealth acquisition, in contrast, is not bound by such limits in its rationality 1258a–1258b).

On the other hand, management essentially has an ethical nature, not only because it is a practice (*praxis*)—an art in Aristotle's words (*Pol.*, 1256b)—but also because it inherently embodies normative elements in its etymological definition.

Indeed, 'management' comes from the Latin 'manus.', which means hand. In the sixteenth century, it meant 'to handle, train, or direct (a horse),' from the verb 'manage,' which refers to 'the handling or training of a horse; horsemanship.' Thus, the term is related to the act of controlling (of a horse). Historically, the modern administration profession originates from cameralism (*kameralismus*) the Prussian administrative system (eighteenth century) characterized by the training of professional administrators for the State and commerce. Cameralist training—which became the basis for the first MBA created in the University of Pennsylvania by Joseph Wharton in 1881—included accounting, economics, law and political philosophy. This educational plan was oriented towards bureaucratic work—a word that originates in *bureaucratie*, from the French bureau, 'office, desk,' and from the ancient Greek *cracia* (*κρατία*), which means—also like *manus*—'government, control or power.'

On the other hand, the term administration—from the Latin *administration*—meaning 'the act of giving or providing.' The verb is composed of *ad*, which means 'to,' and *ministrare*, which refers to 'service or attention.' Therefore, to administer means 'to serve.' In this sense, administrative activity not only refers to the control of productive resources but also, and especially, to the delivery of a specific service, namely, the provision of the productive, organizational, and economic means needed for work and consumption. These aspects are what Finnis (2011a) refers to as 'common stocks' and 'incidents of communal enterprise.' The former pertains to what we commonly refer to as productive resources, capital goods, assets, or products. The latter refers to the policies or decisions that indicate how collaborative processes are carried out within the projects of a common enterprise.

In this sense, a key element of management practice is the controlled provision of resources to enable other workers to fulfill their tasks—as described in Aristotle's terms. Thus, the key element is the provision, not necessarily the property of those means by managers themselves. It is true that the allocation of tasks and the facilitation of collaborative processes—often associated with bureaucratic systems and policies—is justified not only by the need for coordination but also by the principle of private property, which in fact underpins the legitimate power to organize the common stock at will. However, as happens in modern management, non-owner agents have been granted decisional power by owners without management roles in the firm.

Accordingly, management, in its role as providing the conditions for work to take place, assumes akin of a threefold responsibility, which is the need to decide in light of what is fair for investors, workers' resources, and themselves.

3.6.1 *Responsibility to Shareholders: The Context of Corporate Governance*

Modern corporations, as we know them nowadays, emerged more than five centuries ago during the Renaissance in northern Europe, and they were mainly characterized by a separation between management and property.

Indeed, the early version of companies developed as corporations occurred in sixteen century Netherland with the occasion of the emigration of Sephardic Jewish communities from Spain (1492) and Portugal (1495) following their expulsion. This group had extensive experience in managing capital, possibly due to historical restrictions on land ownership and participation in agrarian economies. In Dutch ports, Sephardic merchants applied their expertise and financial resources to support departing commercial ventures. For this, they developed a system where investors could participate in shipping expeditions by contributing a percentage of the ship's departure cost, essentially making each commercial venture a shared investment. This practice laid the groundwork for what would later become the Amsterdam Stock Exchange and, subsequently, the London Stock Exchange, as part of this community moved to England.

This investment practice, however, was eventually done under a formal institution called corporation in Dutch law. Indeed, the Dutch East India Company, along with other national companies that followed, emerged in part due to the growing use of the corporate legal structure—originally corporations where the legal protection to avoid the expropriation from the king. Already codified in the sixteenth century, this structure allowed for the long-term safeguarding of capital, enabling companies to capitalize on commercial opportunities in Asia without needing to find new investors for each expedition. This system protected capital through legal means rather than contractual agreements, which were often difficult to enforce in court or through other authorities. Accordingly, the legal framework of corporations shielded investors from the risk of having their capital seized or expropriated. In a similar manner, commercial corporations safeguard the possibility to withdraw capital, even from investors. The only way to recover the investment was by selling the company's property, that is, shares or stock. In this way, the Dutch East India Company became the first institution capable of securing investor´s capital legally. Investors retained ownership of their shares but could not withdraw the capital used for commercial operations. By 1657, the British East India Company, founded in 1600, had adopted a similar model, securing its capital under the corporate structure.

The creation of the Dutch East India Company in 1602 transformed Amsterdam into a major regional financial hub. Building on the financing practices introduced by Sephardic businessmen, the Dutch Company adopted an ownership structure that featured easily tradable shares. In just a matter of days, interested parties could buy and sell shares, setting the stage for a more fluid financial market. Soon, the Dutch public was engaging in increasingly complex financial transactions, including forwards, futures, options, and bear raids. By the seventeenth century, the practices of the Amsterdam financial market closely resembled those used in today's markets, when corporations had evolved into independent legal entities with their own capacity to act, that is, managed separately from the direct participation of individual investors.

This shift necessitated a series of changes to corporate administrative systems across Europe—the so-called agency—which was characterized by the delegation of authority in directors and agents, the limited liability of corporations, and the creation of the stock markets.

Indeed, the development of the figure of the manager, representative, or agent, allowed investors to delegate their authority to enter contracts on behalf of the company, especially in commercial agreements, where responsibility for an agreement falls on the company, not on the agent himself. This gave way to the constitution of a limited liability corporate entity, that is, protection of the corporation's assets from creditors who provided credit to the owners of the corporation. In addition, protection of the corporation against the principal owners' personal loans, preventing the withdrawal of the investors' capital or a forced liquidation of the corporation. In this vein, access to investments became limited to a secondary share market, transferring related movements to a stock transaction rather than withdrawal of capital. Thus, the corporation always retains its equity, and owners benefit from increased corporate economic value by liquidating their stocks on the stock market.

By the eighteenth century, these commercial corporations not only established international trade relations, but also operated with local administrations in various colonies and continents. They thus required agents to represent the interests of the companies overseas, an administrative scheme that needed to be based on agency duties that, potentially breached, represented for investors what Adam Smith would describe as agency problems between directors—representing the interests of investors—and managers.

Indeed, the management scheme, which separated corporate governance—directors or board members—and agents, gave rise to the problem of managerial duties since long-distance control was practically impossible. Adam Smith first highlighted this situation, calling it the agency problem. Indeed, theoretical engagement with managerial duties in modern organizations dates back to the *Wealth of Nations*, published in 1776. At that point, managerial duties started to be progressively recognized in English commercial law as what later would be defined specifically fiduciary duties, that is, the responsibility of managers to act in good faith and with loyalty.

The Scottish philosopher realized that a new form of productive organization was emerging as a result of international trade characterized by owners with agents acting as administrators abroad. For Smith, agents act according to their own discretion in managerial and commercial matters, which carries the natural risk of them looking out for their own benefit rather than that of the company. Control, in this sense, was hardly impossible, no matter what the contractual requirements were established between agents and investors.

Thus, when the whole operation of early corporations—as well as modern ones—was established under a separation of administration and investment it required—at least according to Smith´s perspective—a strong ethical ground without which it was impossible to understand the nature of corporate relationships, their strengths, and their limitations.

In particular, such ethical ground in the case of managers is established under three essential principles, namely, loyalty, trust, and care. The latter, however, has a different condition, when agents are not necessarily caretakers. It is also known as duties of performance and requires employees to comply with contracts in which their duties are explicitly defined in terms of specific obligations, standards, competence, and diligence in more objective terms (Cooney 2009). Unlike the duty of care that can be set into a contract, the relationships of agency are beyond contractual schemes. In fact, contracts of agency can only stipulate agents to avoid individual interest in their discretionarily (Smith 2002).

3.6.2 Loyalty in Management

The term 'loyalty' has its linguistic root in Ancient Latin *lex* (law) and the Medieval Latin *legalitas* (legal). In the old French (*loialte*) and early Middle English (*leaute*) usages, the signification concerned principles, sentiments, and the status of civil subjects of rulers or governments. Thus, loyalty as a principle meant honesty, commitment, faithfulness, and allegiance. The sentiment of loyalty was associated with devotion or enthusiasm. As a status it meant being faithful, honorable, or law-abiding.

As a sentiment, loyalty has been principally related to its psychological and attitudinal connotations. According to Hart and Thompson (2007), loyalty in a psychological approach has been considered around material or economic goods (transactional loyalty), persons and/or groups (relational loyalty), and causes that bind fidelity (ideological loyalty). On the other hand, loyalty in terms of reputation and, therefore, status, has shown a prominent role not only in civil society, but particularly in markets, as when loyal customers repeatedly buy products or services from a particular business (Hajdin 2005).

As a principle or moral criterion, loyalty signifies a form of commitment. From the Latin *committere*, this concept means 'to give in charge, entrust, to unite, connect, combine; to bring together.' It comes from *com* (together) and *mittō* (to send). As a commitment, loyalty implies a relationship in which a commissioned party acts with a certain degree of autonomy on behalf of an entrusting party. Thus, loyalty befits representation schemes, which are certainly not a relationship in which a party replaces the other. Loyalty, indeed, is proper to the role of trustees acting on behalf of an instructing party. A good or loyal trustee accordingly can be properly defined as an autonomous committed substitute.

Bryant (2004) explains that loyalty applies to what is due or required and specified as an appointed person or cause associated with an entrustor that has the right to be benefited by the former in the specific form of 'acting on behalf.' In this form of representation, trustees are expected to aim for the benefit of their entrustor—as if their benefit were also proper to the same trustee, that is, a community or association based on a shared cause. Indeed, such a common cause is what we can define as the objective of loyalty. The specificity of such a relationship is based on the responsibilities assumed by those who have been given the charge of acting on behalf of another party for the benefit of that same party. According to Kleinig (2007), loyalty is an association attachment that involves a costly commitment to persist or simply not neglect the common interest of two parties. For Melé (2001), loyalty demands dedication to a common goal; a dedication according to which different parties constitute a community or partnership. Loyalty and community—continues Melé—are intrinsically connected: the lack of loyalty deteriorates any community and vice versa. Loyalty is proper to a relationship in which the interest of a first party becomes a common interest when the latter wants that interest for its own value (Allen, 1989). Therefore, loyalty is not oriented towards a person to whom someone can be loyal, but to a cause that involves benefit for that person.[5]

Hence, loyalty-based relationships are those in which one person acts primarily for the benefit of another considering the interests of the beneficiary (as such interests are also their own) and decides so as to serve those interests, even when such actions could imply some costs on the same agent (Smith 2002). However, a relationship of loyalty must not imply for trustees to simply act for an entrustors no matter the consequences for them, accepting as part of the representation scheme the possibility of being harmed because of benefiting an entrustor. Loyalty is neither servility nor domination (Melé, 2001). Loyalty, being a real commitment to an entrustor's interest, implies that trustees must act as if such interests are also a benefit for him/her.

[5] 'Participation in value' and 'commitment' … These notions, as I use them, are rather similar to the existentialist concept of project, as explained by George Kateb, 'Freedom and Worldliness in the Thought of Hannah Arendt' (1977) 5 Political Theory 141 at 153: 'The project is a task without boundaries; one can never say that it is done, yet the whole meaning of it is found in every action done for its sake. ("For the sake of " does not mean "in order to" [Arendt, The Human Condition (Chicago and London: 1958), 154, 156–7]). It is never realized. The fact that I adopt a principle prevents no one else from adopting it; it is inexhaustible... A principle is not a consideration external to the act and reachable by a neutral method'. Of course, as a matter of words, this is precisely not my concept of project (2011a, p.77).

Therefore, loyalty is not a condition for exploitation, but a relation of mutual benefit derived from a single cause.

Even so, avoiding personal harm for a trustee does not imply there is no need to assume risk. On the contrary, being a trustee implies an openness to risking personal benefits as happens with most individual decisions. In other words, the relationship of loyalty does not imply assuming necessary harm or loss, but potential risks in looking for a common interest for both trustees and entrustors. Thus, the relationship of loyalty is proper to a shared enterprise in which different degrees of risk are mutually assumed by trustees and entrustors.

Moreover, according to Melé (2001) being loyal also implies remaining free, reasonable, and responsible, creative, and acting with personal initiative. In fact, such features of a trustees are consistent with the reasons why any entrustor establishes a relationship of loyalty: trustees are in charge of acting on behalf of entrustors because: (i) the former are considered experts (a professional equity or state manager, for instance) and/or (ii) because the entrustor does not have the required time for managing its own capital or state properly (Frankel 1984, 1995). Unreasonable and servile trustees would naturally be incapable of acting in a discretionary manner in their role as experts and dedicated managers.

On the other hand, relationships involving loyalty vary by domain and character. Loyalty pertains to civic[6] or state[7] relationships, among others (Finnis, 2017). This distinction of 'loyalties' is based on the different nature of common causes (for instance, civic or public common causes). Accordingly, agency, which is the specific form of loyalty-based relationship in productive and economic affairs, has its particular cause, which is the economic rent resulting from managing the entrustor's state. Thus, agency refers to the role of trustees participating in a loyalty-based relationship whose common cause is the entrustor's economic benefit potentially obtained from his/her capital or state. According to Frankel (1984, 1995), agents, in this context, are those in the roles of directors, officers, administrators, bailees, or guardians (Frankel 1984, 1995) acting on behalf of stockholders, owners, investors, or employers that, acting on behalf of investors or owners, establish a surrogative relationship of agency with certain employees; any other form of commercial relations, such as those established with clients or lenders, are not based on any form of substitution as in agency, as well in the case of employees discretionary interacting with third parties on behalf of the firm (Cooney 2009).

---

[6] Supervision of the moral-cultural-educational environment is maintained as a very important part of the state's justification for claiming legitimately the loyalty of its decent citizens (Finnis, 2011b, p. 335).

[7] Loyalty is not directly to the law, but to the social order (public good and common good), which requires institutions that, in turn, need legal organization. Therefore, for reasons of loyalty, there is justification for certain forms of civil disobedience (McChrystal 1992, 1998).

3.6.3 In Good Faith or Fiduciary Responsibility

As agency strongly relies in a trustworthy context, the responsibilities borne by agents to entrustors are not only those derived from an agent´s attachment to a common cause—i.e., loyalty—but also a commitment to what sustains such a relationship, that is, trust. This is the specific duty of trust—fiduciary duties or duties of good faith (Gold 2009)—without which the duties of care and loyalty are impracticable. Put differently, the fiduciary duty of trustees not only affects other duties such as the duties of care or loyalty (Johnson 2003), but the fiduciary duty is a duty in itself.

Etymologically 'fiduciary' means 'trust', from the Latin *fides*, that is, confidence or believing. As for agents, this concept implies the importance of fulfilling what is expected of them in their role as trustees, that is, being recognized by entrustors as a loyal agent to the common cause. In other words, agents assume a fiduciary duty not directly in terms of being loyal, but specifically in safeguarding confidence. On the contrary, in a distrusting relationship, any commitment for loyalty borne by agents is futile. For Gold (2009), loyalty needs a fiduciary refrain from lying to shareholders, illegal behavior, breaking commitments, or being involved in misconduct even when these practices can eventually benefit the corporation and shareholders economically. In this vein, DeMott (1988) explains that the fiduciary duty of any agent goes beyond fairness and honesty and is also a matter of doing what is needed to assure confidence between related parties.

According to Eisenberg (2006), fiduciary duties are responsibilities grounded in good faith, which require trustees to act with honesty, sincerity, fidelity to their office, candor, and decency. Decency, in this context, means that trustees must meet the reasonable expectations of society for managers, professionals, or directors. Honesty, on the other hand, refers to adhering to the reasonable standards of fair dealing. Fidelity involves fulfilling the expectations tied to their role in relation to the constituencies with whom the organization typically interacts.

Candor, as part of the duty of good faith, calls for trustees to be frank. In contemporary professional ethics, this is akin to the principle of accountability. Candor requires trustees to avoid misleading statements and to provide necessary information for relevant constituencies to fulfill their obligations.

In essence, breaching any of these aspects of good faith undermines the foundation of trust-based relationships, thereby jeopardizing the potential for sustaining agency relationships. As a result, mistrust and disloyalty, though deeply connected, represent distinct aspects of agency relationships.

Fiduciary managers are individuals entrusted with the administration of a third-party estate for the benefit of those third parties, not for their own gain or the benefit of others. Specifically, this means that an owner cannot instruct fiduciary managers to pursue profits at any cost or for just anyone. In this context, fiduciary relationships involve fiduciary acts carried out on behalf of a beneficiary, with discretion exercised over critical resources belonging to the beneficiary. Thus, the agent's duty, based on the nature of the fiduciary relationship, is to maximize the economic benefit for the beneficiary—the trustee—in exchange for economic compensation, such as professional fees, shared benefits, or both. On the other hand, fiduciary managers are typically sought off for two non-exclusive reasons: either because the agent is an expert in the field or because the beneficiary lacks the time required to manage the administrative tasks effectively.

The economic and commercial performance of fiduciary managers depends on their discretionary decision-making, which distinguishes the fiduciary relationship from other commercial transactions, where results can be stipulated ex ante. Discretion means that fiduciaries make decisions about how to fulfill their duties, but the administrative and strategic choices they make are not formally outlined in ex ante contracts—since no agent can guarantee the future economic success of their actions. Any other business relationship that incorporates some of these conditions, such as those with customers or lenders, does not involve the same form of fiduciary responsibility. In fact, trustees are not guaranteed the highest possible return on the managed assets. Instead, they are expected to act in the best interest of their beneficiaries to achieve optimal outcomes.

3.6.4 Institutional Responsibility: The Provision of Common Stocks for Work

There are two considerations on time that are quite significant for a definition of a firm, namely, when they start and how they perdure over time. The former, indeed, indicates that firms are founded, they are an entrepreneurship, which entails that its future is unpredictable—the terms entrepreneurship and entrepreneur, in fact, were first used to refer to a company in eighteenth-century France by economist Richard Cantillon in his work *Essai sur la nature du commerce en général* published in 1755.

The second defining consideration, perduring over time, has a direct relationship with the role of manager. As mentioned before, management—in light of Aristotle thinking—is a matter of providing the productive, organizational, and economic means necessary for work and consumption. These means are, in the words of Finnis, common stocks or elements of a communal enterprise—essentially, what is required for a shared endeavor to sustain the work of its members over time. In other words, it is about ensuring the sustainability of resources needed for work as a practice to take place.

According to MacIntye's neo-Aristotelianism, this provision of common stocks in a permanent manner is what we can define as the institutional dimension of a firm. From a MacIntyrean approach, collaborative work can be effectively performed in what he describes as institutions, which are necessarily concerned with external goods, and they refer to the procurement and distribution of goods external to practices, some of which are material, such as money, and others, closely related to the material, like power and status. Just as external goods are necessary for practices, institutions are essential for collaborative practices. In other words, no practice can survive for any length of time if not sustained by institutions (MacIntyre 1981).

Accordingly, a firm is not only a community of work, but also an institution as it provides the conditions for collaborative work to take place.

However, the sole provision of the material conditions for work in institutions does not guarantee that work will be performed in an excellent and virtuous manner. The role of the institution is an effective and efficient provision of means for work to take place. However, when firms act as if those means are the ultimate goal of a work, they become corrupted and, particularly, it is possible to sustain that management becomes a corrupted practice, that is, unethical.

In this sense, managers assume responsibility with the institutional dimension of the firm as they care about providing the technical and material condition for good work to take place—i.e., a job that fulfills the four dimensions of a good work, as we have explained before. But this, not sporadically, in the short term, but in a sustainable manner. I the words of Scruton and Finnis (1989, p. 251) "Institutions are characterized by procedures and roles which exist independently of those who make use of them; they have a longevity conferred by the principle of succession to office, and this longevity is something which, in the nature of things, is not shared by an organism. Not every institution has personality, although every institution has some measure of autonomy." For Weber (2009) an organization is a system of continuous, specific and purposeful activities. A corporate organization is an associative social relationship characterized by a managerial staff engaged in said activities with a continuing purpose. This, as we have said, is characterized under the concept of institution.

Accordingly, the institutional nature of a company directly shapes the definition of responsibility for those who manage it. The purpose of management is, in this sense, institutionalization—specifically, the real, coherent, permanent, and unified integration of all aspects of a business that ensure its long-term sustainability. An institution, in this context, becomes the framework that transforms an idea, internal value, or inspiration into a functional and enduring reality. Institutional responsibility, therefore, refers to the duty borne, especially by managers—or any other member of the firm in a secondary manner—to ensure its continuity over time by providing the necessary resources for effective work. In short, it is the responsibility to safeguard and preserve the institution's long-term viability, and, according to this, many negligent practices of managers are in fact a breach of their institutional responsibility when they expose the continuity of the firm.

In particular, managers' institutional responsibility can be outlined in terms of development and growth, autonomy, the long-term, sustainability, and capitalization. Indeed, the specific economic nature of a company—particularly its commercial activities—is possible because of the effective and efficient satisfaction of consumers' needs which, on their part, depends on a productive process. The goal of any firm is to establish this relationship between commerce and production as an autonomous economic cycle, where there is no income outside of the relationship between work-consumption, except for the contribution of capital. In other words, such a relationship between work and profit allows for the material autonomy of a company, that is, its capacity for survival or sustainability, as we shall see below.

However, a company's material autonomy is not its sole purpose. Strictly speaking, if we consider autonomy only from the economic point of view, a company's activity is valued only in terms of economic growth, forgetting the importance of the integral development to which the enterprise must commit. Growth has a strict economic justification in the principles of economy of scale, competition, vertical integration, among others. However, this cannot be justified by itself, but only by the good that these economic improvements allow for the personal development of those who work in the company, of those who benefit from its products and services, and of society at large as well.

3.6.4.1 Human Capital

In line with the need to ensure the long-term performance of the firm and to enhance employees' ability to perform their roles through their work, managers take on the responsibility of increasing the so-called human capital within their company, which reflect the institutional value of the firm's members.

Human capital is rooted in the idea that a company is a social entity that fosters the development of individuals through their work, while simultaneously contributing to societal progress by meeting market needs. Consequently, the company's continuity positively impacts long-term human development, which in turn leads to greater well-being in the workplace.

More specifically, human capital refers to the relationship between capital, labor, business, and wealth. The term was first coined by economist T.W. Schultz (1961), who demonstrated why education should be regarded as an investment with both economic and social benefits. The people who make up a company are considered a form of capital for the organization. This, however, is understood primarily in the context of the relationship between work and wealth. Human capital is a managerial responsibility that links labor with employee development and the overall strength of the institution. Therefore, employees are capital not as tools, but as subjects of development who, through continuous improvement, offer the company opportunities for long-term stability and growth. In this sense, capital is inherently dependent on human labor, not the other way around.

Capital represents the net value of wealth—the value of assets after subtracting liabilities. The term comes from the Latin *capitalis pars debit*, which referred to its use for accounting purposes. Historically, its application in accounting originated in the agricultural economies of ancient and medieval times, where wealth was associated with land and livestock—particularly the heads (capita) of cattle (res).

Wealth, in turn, is capitalized through human labor. Labor transforms wealth into capital, increasing the economic value of wealth (financial capital) or, conversely, rendering untapped wealth (such as property) incapable of being capitalized.

Capitalization requires legal private property, formalizing gross wealth into what is known as working capital. Poverty often arises from the lack of formal recognition of wealth, which is stagnant or unable to be sold due to the absence of property rights. As for labor, it becomes capital through its capacity to capitalize. This can occur in three ways: first when labor aligns the company's resources (structural capital) with market demands through commercial channels (relational capital). Secondly, when labor creates intellectual capital that materializes into available resources for the organization, such as processes, patents, and technological innovations. Finally, when labor generates financial capital, which is liquidity accumulated through labor, thus making financial wealth dependent on human capital.

For each individual in the company to solidify their human capital, employee development must be strengthened across three key areas: technical knowledge, practical knowledge, and virtues. These aspects of personal development positively impact on the company's technical, strategic, and cultural capital.

Through employee development, a company gains various forms of capital, or institutional security, including (1) technical capital (i.e., resources related to technical knowledge), (2) strategic capital (meaning the decision-making structures and the experience and habits that guide those decisions) and cultural capital (a capital existing as shared values and norms within the organization) (*See* Sison 2003).

Finally, it is essential to consider the difference between instrumental and responsible capitalization. Indeed, the capitalization of wealth through labor can be approached in two distinct ways. The first one is instrumental capitalization when labor serves an instrumental role solely focused on generating wealth, making workers functional components of the system. The second one is responsible capitalization, which happens when Labor is prioritized and, accordingly, the firm recognize that capital depends on workers´ capacity and development. In this approach, the company aims to benefit the worker as an individual, strengthening a community of work through participation. In this model, resources are not merely allocated to workers, but workers themselves are responsible for ensuring the development and sustainability of these resources.

Thus, we can conclude that a manager's institutional responsibility, when effectively executed, represents a form of institutional capitalization where work, which is central to the company's sustainability, is regarded as an end in itself

3.6.5 Ethics of the Economic Interest of Managers

As we have noted, managers are expected to act with loyalty, trust, and care. However, the need for managers to receive economic rewards or compensation for their work can sometimes undermine their commitment to these ethical principles. In fact, many well-known scandals in management can be classified as unethical practices where the pursuit of economic gain took precedence over moral responsibilities to shareholders, employees, the institution, or/and society.

However, the problem of unethical approaches from managers to economic interests has been treated in two main lines of thought, namely, agency theory—based on classical Utilitarianism—and virtue ethics, particularly, as the virtue of magnanimity.

3.6.5.1 Agency Theory's Utilitarism

The twentieth century theory of management, and particularly Agency Theory, instead of reinforcing the need for ethical principles, intended to solve the problem only in economic terms. This financial theory, indeed, emerged in this context of corporate separation between control and ownership, postulated that the so-called fiduciary responsibilities can be guaranteed in a correct scheme of economic incentives reflected in the contracts celebrated between owners—*i.e.*, shareholders—and agents.

According to Jensen and Meckling—the most prominent authors in Agency Theory—these incentives translate into performance bonuses, company shares, among others. These, in order to be correct in terms of economic rationality, must represent the best opportunity cost for agents, so they can value the need to stay acting as managers for those shareholders as they would not obtain more benefits in another agency-based relationship. In this way, if incentives are properly placed, the agency relationship is secure, and both parties mutually benefit. On the contrary, when incentives are not properly aligned, the agent may terminate the relationship—in line with opportunity-cost calculations—or so-called agential problems could spring up, that is, decisions and practices in which the agent benefits at the expense of the principal, particularly opportunism.

However, even when these two fundamental elements of the agency relationship—contracts and incentives—are correctly arranged, the risk of opportunism remains, which is why the principal must set up control policies. In the agential scheme, the possibility of opportunism is called moral risk, to which Agency Theory authors add the so-called adverse selection problem. This occurs when, at the time of being interviewed, the agent claims to have competencies and skills that he does not really have and that the principal cannot fully verify.

For Agency Theory—at least in Jensen and Meckling views—an adequate anthropological framework is required for the description of corporate human behavior. For them, human behavior can be understood as based on three universal characteristics: resourcefulness, the ability to evaluate, and the ability to maximize. These characteristics are then integrated into REMM (the Resourceful, Evaluative, Maximizing Model).

This model is based on the following postulates: (1) the individual is interested in various goods (honor, knowledge, interpersonal relationships, status, culture, social norms, honesty, rules of conduct, health, music, art, etc.), which he evaluates, substitutes, and balances according to his preferences. (2) Individual desire is unlimited; the individual prefers a larger quantity of goods than less (and these can range from pieces of art to ethical standards). Moreover, the individual can never be fully satisfied because he will always want more goods, both material and non-material (e.g., companionship, respect, love, fame, immortality). (3) Each individual is a maximizer who seeks the highest possible level of value when satisfying his needs, but, at the same time, he is limited by various factors, such as scarce resources, physical laws, his own knowledge, etc. (4) The individual is resourceful and creative, capable of learning and creating new opportunities.

Finally, Jensen attempts to complement REMM with the Pain Avoidance Model (PAM), adding that it is important to take into account the fact that people are capable of irrational behavior or lack of self-control in difficult situations, which also implies agency costs and requires attention. Agency Theory's approach is not entirely novel in that it updates eighteenth-century utilitarian philosophy, especially as seen in Bentham's ideas.

For Jensen and Meckling, REMM is the basis for the more or less unified development of a theoretical body in the social sciences. According to these authors, some economists, such as Gary Becker, have applied a REMM-like model to other fields of research usually reserved for sociologists. In other words, the authors of AT consider Becker to be a precursor of REMM, not only because his work is based on a model of human action defined by individual calculation, but also because of the broad scope of application that this model has in human affairs.

Becker, in effect, suggests that recognizing the value of the perspective of economic theory for defining behavior also means explicitly assuming an assessment of maximizing behavior, that is, of behavior oriented to utility or wealth, whether at the level of family, business or government. This economic approach is not limited to goods or material desires, not even to the market sphere alone, but to all spheres of action where opportunity costs arise in the use of limited resources. Thus—Becker explains—the economic approach to action is broad and applicable to all human behavior, whether in the market or not, in large or small decisions, for emotional or mechanical ends; be the people involved rich or poor, men or women, intelligent or unintelligent, businessmen or politicians, teachers or students. For Becker, this idea, present in Bentham's philosophy, makes it clear that the calculation of pleasure and pain is applicable to all human behavior and ultimately brings human anthropology into dialogue with scientific systems, in this case, management theory.

In this sense, Becker considers Bentham a precursor to his economic approach to behavior due to its wide scope and calculative nature, which is not limited to monetary issues alone. Indeed, Bentham himself applied his ideas to the study of criminal sanctions, prison system reform, legislative affairs, usury laws, and jurisprudence. As mentioned, the ethics of agency based on individual interest points back to the work of the English philosopher Jeremy Bentham (1748–1832), the precursor of utilitarianism. According to this philosopher, evaluating action must be done in line with instrumental results. Individual utility is, therefore, the driving force of personal action. In this line, Agency Theory, as an explanation of fiduciary action, can be summarized in the following sentence from Bentham:

> That no man of ripe years and of sound mind, acting freely, and with his eyes open, ought to be hindered, with a view to his advantage, from making such bargain, in the way of obtaining money, as he thinks fit: nor, (what is a necessary consequence) anybody hindered from supplying him, upon any terms he thinks proper to accede to. (Bentham 1816)

Here, the social and economic order is seen as derived from an initial value, i.e., economic and individual satisfaction are the driving forces behind human behavior. Thus, the 'virtue' proper to managers is not justice or responsibility, but rather usury and greed.

According to Becker (1976) view, the above is possible because Bentham understands pleasure (which is the same as utility) as a function that appears across the board, in matters as diverse as wealth, friendship, reputation, power, piety, benevolence, knowledge, imagination, hope, etc. Moreover, it is possible in part thanks to the products that can be acquired on the market.

Bentham (1816) presents the idea that usury has a social function, especially in the ability to raise capital as a force to resolve debts and invest, despite existing interest rates, when this allows industry to flourish. Utilitarianism along Bentham's lines justifies the unrestricted use of capital as a formula for everyone to permanently seek productive investments. Indeed, for Bentham, if production and consumption capacities are unlimited, accumulation is positive insofar as it balances out virtue and vice.

As Bentham (1996) notes, the ethics of utilitarianism is defined as the principle that assumes that the most important goal in life is happiness, understood as the presence of pleasure and the absence of pain. When considering that these two feelings naturally determine man's conduct, the criteria for judging whether an action is right relies on determining its contribution (or not) to individual or group happiness. In this way, a community's interest is the sum of its members' interests. Therefore, right actions are those that contribute more to total happiness or pleasure. Bentham claims that his principle of utility is the only standard of right and wrong that everyone across the board can accept. In this regard, Bentham (1996) explains that the legal and moral order of society should take into account the search for the greatest amount of pleasure for the greatest number of people, which corresponds to individual sums of pleasure. This, in Becker's perspective, translates into maximum utility, and is possible in the calculation that allows for the evaluation of scarce means and competing ends.

By the twentieth century, the idea of usury came to be understood rather as 'individual interest' in the sense of greed. To understand economic actions (those happening in corporations by agents and owners), the motivation behind corporate action came to be seen as greed understood as a virtue, in other words, a natural inclination in human beings that motivates individual profit. Well organized greed is now seen as capable of bringing general benefits to the market and society: A sort of ethical selfishness that we can see in some of Milton Friedman's ideas, as we will mention later.

Some economists began to widely promote the 'virtue' of greed during the twentieth century, based on the assumption that Adam Smith gave it a new sense of dignity (Wight 2005). Kenneth Arrow and Frank Hahn, in their book *General Competitive Analysis* (1971), implicitly assume that individual interest is greed that enables economic efficiency. Williams (1999), for example, writes that free markets, private property, free trade, and greed cause better outcomes most of the time and under almost all conditions. While this ethical egoism flourished in American culture, its relation to Smith is misguided. Smith himself broadly rejects selfishness (and usury), although Smith does morally accept self-interest (or self-love) as a basis for good business activity. Thus, Smith distinguishes between greed and selfishness, on the one hand, and prudence and virtuous self-interest on the other: Smith considers people to be social animals and, at the same time, to have a limited capacity for rationality and an outsized capacity for heroism.

Adam Smith was aware of the negative social consequences of greed: "To be anxious, or to be laying a plot either to gain or to save a single shilling, would degrade the most vulgar tradesman in the opinion of all his neighbors" (Smith 1981, p. 173). Consequently, Smith felt that behavior was rightly constrained by social networks, reinforced through the sympathetic responses of fellow citizens. This framework affirms the role of fairness and respect for justice, which implies appropriate moral consideration for our treatment of others, enforced by impartial bystanders, in addition to laws. It is not difficult to imagine other circumstances in which greed diminishes economic benefits. Indeed, greed implies a kind of self-interest that disregards the rights and interests of others (unless something is gained by considering them). Thus, interest in others, too, is a form of self-interest, but only insofar as the good of others is at the same time also considered to be one's own good.

In this sense, the idea that 'those who seek the common good are merely being selfish like everyone else, just doing so behind a facade of altruism,' is not really an argument that can explain all the circumstances in which a person acts with a sense of community. Some indeed bear selfish intentions, but there are also those who consider the good of others as their own. In this sense, they value their own good because it is good, rather than because it is theirs. Such a view cannot be explained by the individualist perspective as anything other than an error.[8]

To put it another way, economic transactions between greedy individuals may involve higher transaction costs when it comes to monitoring and enforcing contracts. Therefore, trust and social capital are, as Fukuyama (1999) explains, a form of competitive advantage.

In short, the whole tradition that goes from Bentham through Becker and takes shape in Agency Theory is nothing more than a utilitarian philosophy that seeks to explain and encourage agents' behavior in favor of shareholders on the basis that human action is good when it fulfills its original calling, namely the selfish pursuit of individual interest. Within this conceptual framework, the search for wealth as the natural motivation behind human behavior can be adjusted and oriented in favor of the cooperative relationship between agents and principals. Economic relations—as these authors assume—will thereby be of maximum benefit for all. Accordingly, greed is nothing more than an inordinate appetite for wealth. Avarice, for its part, encourages desire for things beyond what is necessary, especially material goods that one already possesses. Usury, finally, is the gain, fruits, utility or increase that is taken from something, especially when it is excessive. It is a kind of illicit interest taken for money or goods in a mutual or loan contract. It can be understood as an unjust commercial relationship in which profit is obtained at the expense of another,

[8] It is also important to note that more empirical research has shown that selfishness and greed do not necessarily bring superior benefits to society or to individuals acting as free riders. Thus, for example, the so-called Ultimatum Game has shown that greedy individuals (those who do not consider basic fairness when it comes to economic distribution) tend to gain less than others in negotiations. One cause of this phenomenon may be the fact that greed can arouse feelings of malevolence, causing players to put aside their own interests in pursuit of punishing another player.

especially when the counterparty is in a situation of need. In short, definitionally, these three elements cannot play a positive role in outlining good fiduciary management.

3.6.5.2 Redefining Ambition as a Virtue

Contrary to this utilitarian view, it is important to take into account that responsibility cannot be achieved on the basis of external incentives alone (such as contracts or economic benefits) but rather grows in the context of personal commitment backed by one's own professional principles. Incentives are different from responsibility since the former only consider what might externally motivate human action. The latter, on the other hand, takes on a more deeply rooted view of people's behavior because it arises from the sphere of subjectivity, from the inner self. This is why a strategy oriented towards a culture of responsibility is better than simply implementing incentives that ultimately corrupt internal motivation in terms of greed, avarice, and/or usury.

Accordingly, the sole use of economic or external incentives is far from strengthening any fiduciary responsibility or loyalty of board members and managers, because is—according to MacIntyre—an individualistic value of wealth accumulation has affected all areas of human life; personal work included. According to him, "the tradition of the virtues is at variance with central features of the modern economic order and more especially its individualism, its acquisitiveness and its elevation of the values of the market to a central social place" (MacIntyre 1981, p. 254). Hence, modern market economies are still highly associated with a corrupt form of society dominated by "markets, factories and bureaucracies over individuals, in which a narrative unity of human life and of a practice with internal goods is excluded" (MacIntyre 1981, p. 228). Among other features, MacIntyre sustains that modern markets are corrupted because of the overestimation of acquisitiveness—the Aristotelian vice of *pleonexia*—and its role in what he calls the *ethics-of-the-market* (MacIntyre 2016). This means that "acquisitiveness is not only a character trait, but also a duty. Individuals owe it to those who have invested in their enterprises to maximize the return on their investments" (2016, p. 127–128).

Therefore—and contrary to an individualistic approach to managers´ ethics—we must understand how to increase agents' commitment based on their own or internal motivations, that is, in terms of virtues. To do this, we follow the structure of classical ethics and consider two fundamental aspects of non-individualistic personal motivation, namely virtue and the common good, in the definition of healthy professional ambition of managers or magnanimity.

Following Aristotle and Thomas Aquinas, we argue that magnanimity is a managerial virtue proper to those who seek to grow professionally by contributing to the company they work for. In other words, if ambition is ethically good, it is defined as magnanimity, and it is usually present in good team players. According to Aquinas (*S.Th.*, II-II, q. 129, q. 2), magnanimity is the virtue that characterizes those who aim to achieve objectives and goals that are worthy of recognition (honor). These achievements are seen as valuable, however, not because of the mere recognition they bring, but for their innate value.

Moreover, ambition understood as magnanimity implies a pursuit of achievements that are valuable in and of themselves, and that are difficult to attain. In this regard, Thomas Aquinas explains that magnanimity also implies a degree of uncertainty (hope) because the difficulty of the activity makes a successful outcome uncertain. Therefore, magnanimity as a virtue implies a certain self-assurance, a certain tranquility or fortitude in the face of uncertainty, that is far from blind confidence in one's own abilities.

Yet economic goods have an instrumental value for the magnanimous person since no worthwhile human activity can be achieved without the necessary resources. It is important, Thomas Aquinas stresses that motivation be the achievement that requires money, rather than money itself. Indeed, for Aristotle, those who are magnanimous are moderate in regard to riches.

To this, we can add that business-related achievements can produce profitability, but they should not directly seek out profitability, but rather the goal or objective that leads to profitability. As a virtue, it is important to take into account that magnanimity, while it is considered good ambition in pursuit of hard-to-obtain achievements, cannot ignore the way in which said achievements are attained. In this sense, ambition may be licit regarding an end, but illicit in regard to the means to achieve it.

Therefore, magnanimity and individualism are opposites because, without considering the common good, ambition cannot be truly good. Thus, an achievement is genuinely valuable as a common good, not simply as a good to be enjoyed individually. Thus, magnanimity and individualism are incompatible, for without consideration of the common good, ambition cannot be truly good. Hence, we speak more correctly of personal, not individual, goods, when the 'personal' character incorporates the sense of community. All personal goods must indeed be understood, valued and pursued in an articulated way with others, in the context of a common good.

Thus, everything I want for myself, in all its dimensions, must be wanted in such a way that it benefits other people. For example, when I seek to develop my career, I must do so in consideration of others' careers, which means promoting others or, at the very least, not harming them in their professional development. Therefore, the first criterion of fiduciary responsibility is to seek for others what I want for myself. This is a criterion related to justice and the common good.

Therefore, self-interest in terms of virtue is not individualism, but rather the will to achieve valuable goals that are, at the same time, both personal and common good.

3.6.6 Practices Associated with Management and Agency Ethics

These pertain to the practices associated with managers and their relationships with investors, markets, and shareholders, and may also include controversial practices among shareholders.

3.6.6.1 Opportunism and Favoritism

In the context of separation between shareholders, directors, and managers, two distortions in the corporate relationship can appear, namely favoritism and opportunism. The former, favoritism or unilateralism refers to behavior on the part of a manager that benefits shareholders at the expense of other third parties usually interacting with the corporation. In this regard, favoritism arises when benefiting a shareholder damages the patrimony or economic expectations of third parties linked to the company's activity, for example, the State. Opportunistic behavior, on the other hand, refers, instead, to actions on the part of a controller who seeks to benefit himself at the shareholder's expense.

3.6.6.2 Insider Trading

This is the (frequently illegal) practice in which a member of the company uses insider information for his own benefit or for the benefit of a third party, especially as related to stock transactions. The information used in this practice is called insider information. As indicated, the use of this information is usually punishable by law.

3.6.6.3 Shareholders Conflict

Modern corporations developed during the twentieth century presented some additional challenges, especially related to structure and long term that caused potential conflict between investors. Indeed, the ownership structure of corporations allows for a first distinction to be made between those with diluted ownership among shareholders, which in turn have a lesser or very passive relationship with corporate governance and, on the other hand, those with concentrated equity ownership where shareholders have a preponderant and present role in corporate governance.

As Rojas (2017) explains, the former tends to establish short-term goals, value results in that context, and are more subject to the pressures of sustained quarterly results, which in turn can negatively impact long-term decisions. Conversely, corporations with concentrated ownership are usually controlled by a founder (e.g., Facebook), an investment fund, or hedge fund activists, who imprint an ideological or political leaning on the corporation (e.g., Patagonia). This type of structure favors the long term and tends to be less dependent on quarterly results.

3.6.6.4 Fostering Shareholder Primacy

The ethics of corporate governance aims to reach an understanding of the dilemma between shareholders and managers depending on the ownership structure. In one case, this happens in corporations in which shares´ property is quite diluted, and no shareholder has a majority. In such a case, shareholders do not take priority over controllers—*i.e.,* board directors and managers. This position is called managerialism and is characterized by giving greater importance to managers' activity, particularly regarding decision-making.

Such a perspective can materialize in companies where ownership is highly diluted, there are no majority shareholders, and management decisions do not really have a counterpart. In this sense, the corporation is understood as a bureaucratic hierarchy managed by managers, where the directors are somewhat symbolic figures and shareholders are a minority group. Thus, managers herein are autonomous and free to pursue the objectives they set for themselves (Bainbridge 1998).

Conversely, when priority is given to shareholders, especially in concentrated ownership structures, it is called shareholder primacy. In this case, the shareholder is considered the 'owner' of the company and, consequently, directors and managers are nothing more than employees of the owner, with a duty to act in the owner's best interest. As Bainbridge (1998) explains, this is partly due to the influence of agency theory and establishes that directors or managers are just agents of shareholders with fiduciary responsibilities aimed mainly at maximizing share value.

3.6.6.5 Corporate Activism

Since the seventeenth century, corporations have represented one of the most important institutions of modern capitalism and they rely on investors who seek to make their capital profitable. In line with the axiological nature of corporations, many investors, especially since the twentieth century, seek to control or influence corporate governance in order to guide companies. This, in fact, is done both in terms of investors' own interests, as well as to promote higher standards in terms of the social impact of productive and commercial activities.

These types of investors are often called activists or dissidents. They engage in politics (in the partisan sense, as congressmen or lobbyists) to widen their sphere of influence over boards or management through campaigns, contacts, media, legal means, etc. This type of activist investor has, in general, two types of objectives, namely individual interest (corporate riders) and social interest.

The first historical record of this type of activism appears precisely with the Dutch East India Company. In 1622, a group of minority shareholders published pamphlets to bring to light a series of conflicts of interest that the company's directors had so that, under social pressure, the system for electing directors would be reformulated (likely in their favor).

Today, individual activists are usually understood to orient their participation in companies toward increasing their wealth. Usually, this strategy is called ‘asset stripping,’ which refers to acquiring a controlling interest in the shares of a company by means of an out-of-pocket amount or putting up collateral and then requesting financing from the bank by offering the newly acquired company assets as collateral.

The practices of corporate riders involve, among other negative social effects, a progressive and extensive deterioration of the real value of companies, but in favor of their own equity, illicitly diverting debt responsibility or fiduciary duty. In other words, a debt fraud that, at the same time, damages companies by producing a credit collapse (financial clout).

Another type of individual investor uses a lobbying campaign rather than acquiring a dominant position in the company through debt fraud. This type of investor—who appeared on Wall Street in the 1980s and 1990s—is not so much a matter of seeking a majority stake, but rather a greater degree of influence.

Linked to investment funds, these activists look for weaknesses in a company’s strategy and operation. Under these arguments, they manage to convince other shareholders that a change in corporate governance is necessary, and naturally, they present themselves as the agents of that change.

In this way, with little capital available, these shareholders invest in various small or medium-sized companies, eventually creating a holding company in which they achieve a dominant position. This strategy, as in the case of Enron, can have an impact on the valuation of companies when they appear to be investing in the acquisition of new businesses, but in reality, it is only a network of related businesses. All these strategies, as shown, can be categorized to a greater or lesser extent as opportunistic and have negative impacts on society and the market.

Since the 2008 subprime crisis, there has been an increase in shareholder activism in favor of better corporate standards. This means that, through lobbying-like activity or simply by acquiring a majority stake, some shareholders have integrated into various companies with the aim of improving their social, not just financial, impact.

These practices can be broadly classified into three types. The first involves value-based investors—those with a strong sense of corporate and social responsibility, often grounded in ethical or religious principles that shape their investment choices. While they do not necessarily avoid companies with poor moral performance, their investments in such firms are typically aimed at improving practices or redefining corporate purpose.

The second type comprises value-seeking investors, who target markets or industries that promote sustainability, thereby achieving stronger performance due to the high social regard (or ‘social license’) such companies enjoy. A prominent example is investment in renewable energy.

The third type consists of value-enhancing investors—those who, by virtue of holding a dominant position or exerting significant influence over corporate governance, actively work to improve standards in social, administrative, legal, and environmental domains. In this sense, they act as activists motivated by the ethical enhancement of corporations.

References

Al Halbusi, H., P. Ruiz-Palomino, J. Linuesa-Langreo, and G. Scalzo. 2025. Ethical leadership as a driver of supervisor technical and social effectiveness: A triple helix for cultivating employees' sense of purpose. *Business Ethics, the Environment & Responsibility* 34 (4): 2013–2033.

Aquinas, T. 1274/1947. Summa theologica (fathers of the English Dominican Province, Trans.). New York: Benziger Brothers.

Aquinas, T. 1268/1952. Quaestiones disputate de potentia Dei. https://www.corpusthomisticum.org/iopera.html. Accessed 22 August 2025.

Aquinas, T. 1256/1980. Tomás de, In III Sententiarum, S. Thomas Aquinatis Opera Omnia, 1, In Quattuor Libros Sententiatorum. https://www.corpusthomisticum.org/iopera.html. Accessed 22 August 2025.

Arendt, H. 1958. What was authority? *NOMOS: American Society for Political and Legal Philosophy* 1:81.

Argandoña, A. 1998. The stakeholder theory and the common good. *Journal of business ethics* 17 (9): 1093–1102.

Aristotle. 1995a. Nicomachean ethics. In *The complete works of Aristotle*, ed. J. Barnes, vol. 2. Princeton: Princeton University Press.

Aristotle. 1995b. Rhetoric. In *The complete works of Aristotle*, ed. J. Barnes, vol. 2. Princeton: Princeton University Press.

Aristotle. 1995c. Physics. In *The complete works of Aristotle*, ed. J. Barnes, vol. 1. Princeton: Princeton University Press.

Arrow, K. J., and F. Hahn. 1971. General competitive analysis. In *Advanced textbooks in economics*. Amsterdam: North Holland.

Bainbridge, S. M. 1998. Corporate decision making and the moral rights of employees: Participatory management and natural law. *Villanova Law Review*, 43-741.

Becker, G. 1976. *The economic approach to human behavior*. Chicago: The University of Chicago Press.

Bentham, J. 1816. *Defense of usury*. London: Payne and Foss.

Bentham, J. [1789 (1996)]. *An introduction to the principles of morals and legislation*. Oxford: Oxford University Press.

Breen, K. 2012. Production and productive reason. *New Political Economy* 17 (5): 611–632. https://doi.org/10.1080/13563467.2012.656082.

Ceva, E., and M. Bocchiola. 2020. Theories of whistleblowing. *Philosophy Compass* 15 (1): e12642.

Ciardi, L., and G. Scalzo. 2022. Power, authority, and leadership: A proposal for organizational theory in the post-bureaucratic era. In *Philosophy and business ethics: Organizations, CSR and moral practice*, 569–596. Cham: Springer International Publishing.

Cooney, L. L. 2009. Employee fiduciary duties: one size does not fit all. *Mississippi Law Journal* 79: 853.

de Souza Salles, S., C. F. G. C. da Silveira, and G. Longo. 2022. Dignity is said in many ways: a re-reading based on Thomas Aquinas. *Conhecimento & Diversidade* 14 (34): 271–286.

DeMott, D. A. 1988. Beyond metaphor: An analysis of fiduciary obligation. *Duke Law Journal* 879–924.

Eisenberg, M. A. 2006. The duty of good faith in corporate law. *Delaware Journal of Corporate Law* 31, 1–38.
Finnis, J. 1998. *Aquinas: Moral, political, and legal theory*. Oxford: Oxford University Press.
Finnis, J. 2011a. *Natural law and natural rights*, 2nd ed. Oxford: Oxford University Press.
Finnis, J. 2011b. *Human rights and common good: Collected essays*, vol. 3. Oxford: Oxford University Press.
Frankel, T. 1984. Corporate directors' duty of care: The American Law Institute's project on corporate governance. *The George Washington Law Review* 52:705.
Frankel, T. 1995. Fiduciary duties as default rules. *Oregon Law Review* 74:1209.
Frankl, V. E. 1985. *Man's search for meaning*. New York: Simon and Schuster.
Fukuyama, F. 1999. *Social capital and civil society*. George Mason University, Fairfax: The Institute of Public Policy.
Gomberg, P. 2016. Why distributive justice is impossible but contributive justice would work. *Science & Society* 80 (1): 31 55.
Gómez-Lobo, A. 1989. *La Ética de Sócrates*. Santiago de Chile: Editorial Universitaria.
Harrison, J. S., and R. E. Freeman. 2004. Is organizational democracy worth the effort? *Academy of Management Executive* 18 (3): 49–53. https://doi.org/10.5465/ame.2004.14776168.
Hajdin, M. 2005. Employee loyalty: An examination. *Journal of Business Ethics* 59: 259–280.
Hart, D. W., and J. A. Thompson. 2007. Untangling employee loyalty: A psychological contract perspective. *Business Ethics Quarterly* 17 (2): 297–323.
International Labour Organization (ILO). 1958. *Discrimination (employment and occupation) convention, C111*. Geneva: ILO.
Johnson, L. 2003. After Enron: remembering loyalty discourse in corporate law. *Delaware Journal of Corporate Law* 28:27.
Kerr, J. L. 2004. The limits of organizational democracy. *Academy of Management Perspectives* 18 (3): 81–95.
Llano, C. 1982. *Análisis de la acción directiva*. Mexico: Limusa.
Llano, C. 2000. *La Amistad en la empresa*. Mexico: Editorial FCE.
MacIntyre, A. 1981. *After virtue*. Notre Dame: University of Notre Dame Press.
MacIntyre, A. 1985. Rights, practices and marxism: Reply to six critics. *Analyse und Kritik* 7 (2): 234–248.
MacIntyre, A. 1988. *Whose justice? Which rationality?* Notre Dame: University of Notre Dame Press.
MacIntyre, A. 2016. *Ethics in the conflicts of modernity: An essay on desire, practical reasoning, and narrative*. Cambridge: Cambridge University Press.
McChrystal, M. K. 1992. Lawyers and loyalty. *William and Mary Law Review* 33 (2): 367–427.
McChrystal, M. K. 1998. Professional loyalties: A response to John Kleinig's account. *American Philosophical Association Newsletter on Philosophy and Law* 98 (1): 83–90.
Moriarty, J. 2012. Justice in compensation: a defense. *Business Ethics: A European Review* 21 (1): 64–76.
Melé, D. 2001. Loyalty in business: Subversive doctrine or real need? *Business Ethics Quarterly* 11 (1): 11–26. https://doi.org/10.2307/3857412.
Melé, D. 2012. The firm as a community of persons: A pillar of humanistic business ethos, *Journal of Business Ehics* 106 (1): 89–101. https://doi.org/10.1007/s10551-011-1051-2.
Moriarty, J. 2014. Compensation ethics and organizational commitment. *Business Ethics Quarterly* 24 (1): 31–53.
Mulgan, R. 1999. The role of friendship in Aristotle's political theory. *Critical Review of International Social and Political Philosophy* 2 (4): 15–32.
Murcio, R., and G. Scalzo. 2021. Person-centered leadership: The practical idea as a dynamic principle for ethical leadership. *Frontiers in Psychology* 12, 708849.
Nagel, T. 1972. Aristotle on eudaimonia. In *Phronesis*, 252–259. London: Brill.
Nonaka, I., R. Toyama, and T. Hirata. 2008. *Managing flow: A process theory of the knowledge-based firm*. New York: Palgrave Macmillan.

Pinto, J., and G. R. Scalzo. 2024. How much (more) is fair to pay? The ethics of merit in assessing poverty wages in organizations. *International Journal of Ethics and Systems*. https://doi.org/10.1108/IJOES-06-2023-0132.

Pinto, J., G. Scalzo, and I. Ferrero. 2022. Three rival versions of work and technology: Smith, Marx, and MacIntyre in discussion (coauthored with Javier Pinto). In *Philosophy for Business Ethics*, ed. G. Faldetta, E. Mollona, and M. Pellegrini, 275–199. Palgrave. https://doi.org/10.1007/978-3-030-97106-9_11.

Pinto-Garay, J., G. Scalzo, and I. Ferrero. 2020. Autonomy and subordination: Virtuous work in light of Aristotelian practical knowledge in organizational theory. *Business and Professional Ethics Journal* 40 (1): 47–80. https://doi.org/10.5840/bpej20201227104.

Pinto-Garay, J., G. Scalzo, and C. R. Lluesma. 2022. The narrative dimension of productive work: Craftsmanship and collegiality in the quest for excellence in modern productivity. *Philosophy of Management* 21 (2): 245–264. https://doi.org/10.1007/s40926-021-00188-8.

Redín, D. M., G. Cabaleiro-Cerviño, I. Rodriguez-Carreño, and G. Scalzo. 2023. Innovation as a practice: Why automation will not kill innovation. *Frontiers in Psychology* 13: 1045508.

Robbins, L. P., and W. B. Deane. 1986. The corporate ombuds: A new approach to conflict management. *Negotiation Journal* 2 (2): 195–205.

Rojas, C. R. 2017. *Eclipse of the public corporation revisited: Concentrated equity ownership theory*. Oxford Business Law Blog. https://www.law.ox.ac.uk/business-law-blog/blog/2017/03/eclipse-public-corporation-revisited-concentrated-equity-ownership.

Rosenberg, N. 1965. Adam Smith on the division of labor: Two views or one? *Economica* 32 (126): 127–139. https://doi.org/10.2307/2552796.

Sayer, A. 2009. Contributive justice and meaningful work. *Res Publica* 15 (1): 1–16. https://doi.org/10.1007/s11158-008-9070-1.

Scalzo, G. R., and H. Alford. 2016. Prudence as part of a worldview: Historical and conceptual dimensions. In *The challenges of capitalism for virtue ethics and the common good*, 254–271. Cheltenham: Edward Elgar Publishing.

Scalzo, G., K. Akrivou, and M. J. Fernandez Gonzalez. 2023. A personalist approach to business ethics: New perspectives for virtue ethics and servant leadership. *Business Ethics, the Environment & Responsibility* 32, 145–158.

Schultz, T. W. 1961. Investment in human capital. *The American Economic Review* 51 (1): 1–17.

Scruton, R., and J. Finnis. 1989. Corporate persons. *Proceedings of the Aristotelian Society, Supplementary Volumes* 63:239–274.

Sison, A. J. G. 2003. *The moral capital of leaders: Why virtue matters*. Cheltenham: Edward Elgar Publishing.

Sison, A. J. G., and J. Fontrodona. 2012. The common good of the firm in the Aristotelian-Thomistic tradition. *Business Ethics Quarterly* 22 (2): 211–246. https://doi.org/10.5840/beq201222211.

Sison, A. J. G., and J. Fontrodona. 2013. Participating in the common good of the firm. *Journal of Business Ethics* 113 (4): 611–625. https://doi.org/10.1007/s10551-013-1678-7.

Sison, A. J. G., I. Ferrero, and G. Guitián. 2016. Human dignity and the dignity of work: Insights from Catholic social teaching. *Business Ethics Quarterly* 26 (4): 503–528. https://doi.org/10.1017/beq.2016.42.

Smith, A. (1776/1981). *An inquiry into the nature and causes of the wealth of nations*. Indianapolis: Liberty Fund.

Smith, D. G. 2002. The critical resource theory of fiduciary duty. *Vanderbilt Law Review* 55:1399.

Taylor, F. W. 1911. *The principles of scientific management*. New York: Harper & Brothers.

Thompson, D. F. 2017. Understanding financial conflicts of interest. In *Research ethics*, 505–508. London: Routledge.

Weber, M. 2009. *The theory of social and economic organization*. New York: Simon and Schuster.

Wight, J. B. 2005. Adam Smith and greed. *Journal of Private Enterprise* 21 (1): 46.

Williams, W. 1999. *The virtue of greed in promoting public good*. Richmond Times-Dispatch, October 6, A15.

Chapter 4
Commercial Ethics

Abstract This chapter develops a neo-Aristotelian account of commercial ethics, treating market relationships as commutative interactions guided by mutual benefit, proportionality, freewill, and goodwill. It clarifies just commercial exchanges—from product features and truthful information to fair pricing and client suitability—and frames markets as communities ordered to the common good. It argues that fairness requires proportional gains and attention to consumer sustainability (i.e.: consent alone is insufficient) and outlines criteria—common benefit, proportionality, just price as benchmark, degrees of cooperation, and double-effect for collateral impacts. It links these principles to shared value and the value chain: better products and access, circular productivity, and local clusters that strengthen cooperation. It also contrasts responsible conduct with recurring abuses, including predatory lending, price discrimination akin to usury, pyramid and Ponzi schemes, misrepresentation, and other deceptive bad practices in commerce. Conceiving markets as social communities and exchanges as matters of commutative justice, it offers a practical framework that integrates profitability with human flourishing and the common good. It underscores the social nature of exchange, where benevolence toward counterparties strengthens sustainable, long term relationships and responsible pricing and disclosure.

Keywords Commercial ethics · Commutative justice · Mutual benefit and proportionality · Just price · Consumer sustainability · Shared value · Market as community · Common good

Different from organizational stakeholders are those with whom the firm establishes commercial relationships. These are commercial stakeholders. On a neo-Aristotelian approach, such relationships are commutative interactions, and the corresponding criterion of fairness is given by the mutual benefit, not merit. Following Aristotle, Aquinas states that commutative justice takes place through interchanges and trade-offs between two persons (Aquinas, *S.Th.*, II-II, q.61, a.1). Commutative justice is, thus, a form of justice that applies to trading, usufruct, money lending, deposits, or

J. Pinto-Garay and G. Scalzo, *Justice and Corporate Excellence*, Ethical Economy 72,
https://doi.org/10.1007/978-3-032-12189-9_4

rent (Aristotle, *NE*, 1995a 1131a), i.e., in all actions that manifest the commercial exchange (Aquinas, *S.Th.*, II-II, q.61, a.2).

Accordingly, fair transactions with commercial stakeholder are a function of freewill, proportionality, and goodwill—a view importantly different from Kantian Capitalism in which, as explained above, normative standards are associated with economic practices by means of integrating property rights, freedom, and voluntary action. For virtue ethics, on the contrary, fairness with commercial stakeholders requires that commercial activity be just or fair not simply by being associated with voluntary engagement under a universalizable moral rule but by displaying proportionality in free exchange, that is, accepting deals in which both parties freely attain specific benefits through the transaction (Finnis 2011a). Freewill and proportionality are, therefore, basic features of commutative justice.

However, the scope of commercial justice is not limited to satisfying these two moral standards. Indeed, fair transactions also require the will to support personal thriving and well-being of the related constituencies, that is, intending securement of the common good in market practices. Therefore, commerce must integrate not only the satisfaction of personal needs, but also an intention to benefit those who engage in commerce. This intention of providing to commercial constituencies what they would need to improve their living is known as goodwill or benevolence (Pinto et al. 2021). This is a virtue associated not with philanthropy but the importance of (1) doing good to market agents within the context of profitable interchange and (2) intending to keep commercial agents sustainable—which involves, in finance theory, say, promoting the so-called household finance in market practices (Beshears et al. 2018).

Furthermore, from a neo-Aristotelian approach, transactions and markets are essentially a matter of social interaction. Therefore, the ambit of commercial activities, the market, is in fact a form of what Aristotle calls society. To see why, consider Aristotle´s explanation of the division of labor (Aristotle, *NE*, 1133a) and its importance for satisfying a wide range of needs that households cannot satisfy by themselves (Aristotle, *Pol.*, 1252b). For Aristotle, society exists to the extent of personal interaction is based on a proportional justice that makes possible the cooperation between those who have different occupations or activities. Finnis accordingly explains that Aristotle sees that things will be better for everyone if there is a division of labor between families, specialization, technology, joint or cooperative enterprises in production and marketing, and a market and medium of exchange—in sum, an economy that is more than domestic, but social and based on the division of labor (Pinto-Garay 2015). The physician, says Aristotle, does not associate with another doctor to exchange the product of their service, but does so with the farmer and people of other professions or trades. Additionally, products and services must be interchanged in a proportional manner, and such proportionality is possible when interchange is mediated by currencies or other instruments of economic value (Aristotle, *NE*, 1133b). In this way, the ambit of society is intrinsically associated with commerce since society emerges precisely from the division of labor and commercial exchange.

4.1 Justice in Acquisitions and Exchange

From an Aristotelian perspective, an exchange theory is inherently linked to a theory of justice, and by extension, to the flourishing of communities. In Aristotelian terms, without exchange, we would neither recognize nor participate in the virtues that our community deems exemplary of the human being, as we would belong to no community at all (Koehn 1992).

In a similar vein, MacIntyre contends that market practices must be understood in terms of fairness and the common good. Like Aristotle, MacIntyre adheres to the criterion of justice, positing that acquisitiveness and exchange are fundamentally situated within a framework of cooperation and the common good. In other words, commercial acquisitions and market consumption can be virtuous when rooted in the virtue of justice. Accordingly, engaging in commercial exchange is morally acceptable only when profit is not the primary objective, but only an instrumental one. When profits become the principal justification for economic activity, commerce itself becomes morally flawed, as profits, when treated as a good in and of themselves, are unjust (MacIntyre 1981, 2016). Thus, from an Aristotelian standpoint, any overemphasis on profits above other goods constitutes an act of injustice.

According to Pinto-Garay et al. (2023), MacIntyre's neo-Aristotelianism offers a critical assessment of what he refers to as 'market societies', which he contrasts with a common good economy. This critique not only challenges the view that the sole economic purpose of commercial practices is the pursuit of profits but also provides a more nuanced conceptual understanding of the roots of a highly developed ethics of the market. The so-called *ethics of the market*—as MacIntyre continues—incorporates Aristotle's concept of *pleonexia*, an unnatural form of *chrematistics*, which indirectly points to Aristotle's notion of a natural sense of acquisitiveness (2016). According to Balot (2001), "Aristotle had argued that excessive greed for money, which drives the 'unnatural' mode of acquisition, is rooted in an improper conception of the good life (Aristotle, *NE*, 1257b40-1258a6) … *pleonexia*, namely the immoderate desire for material goods…", that is, a violation of justice. Therefore, Aristotelian *pleonexia* is more closely associated with an imbalanced desire for goods merely for the sake of possessing them. In MacIntyre's terms, this reflects a vicious tendency to acquire instrumental goods, which leads to an increasing emphasis on acquisitiveness and the centrality of market values in society (MacIntyre 1981). Accordingly, the market dynamic that harms personal development and the common good is found in so-called consumption-driven practices, characterized by an excessive focus on consumption. This occurs when a good life—what we might call prosperity—is mistakenly equated with commercial activity and a consumerist lifestyle.

Moreover, the unjust practice of *pleonexia*—or greed—not only underlies deception and misrepresentation, but also usury, which involves the acquisition of economic rent at the expense of another person. Usury refers to the practice of seeking profit in a business transaction where the counterparty does not genuinely benefit but instead is harmed—in the short or long term—or gains nothing. Thus, usury occurs in a commercial relationship where the mutual benefit of each party—an essential criterion of commutative justice, as we have explained before—is absent.

In this context, there are several reasons why the disadvantaged party may willingly engage in usurious deals. Often, individuals accept such terms out of necessity or ignorance. In cases of necessity, a person may agree to unfavorable commercial or economic conditions because they have no other viable option to address a pressing issue. For example, usurious loans may be offered to individuals facing personal or family health emergencies, leaving them with no choice but to accept the unfavorable terms. In this vein, Aquinas (*S.Th.*, II-II, q.77, a.1) explains that a case of usury happens "... when a man has great need of a certain thing, while another man will suffer if he be without it ... He who gives usury does not give it voluntarily simply, but under a certain necessity, in so far as he needs to borrow money which the owner is unwilling to lend without usury."

In terms of ignorance, people may voluntarily accept unfair business terms due to financial illiteracy. It is common, for instance, for the most economically vulnerable segments of the population to accept loans with high interest rates and rigid repayment terms, not fully realizing that these conditions will lead them into financial distress and potentially insolvency.

As seen in both cases of extreme necessity and ignorance, even if the affected party voluntarily accepts the terms of the deal, it doesn't automatically make the transaction fair. The mere willingness of a party to engage in a business relationship does not justify any form of transaction, even when all terms are disclosed. Thus, the issue with usury is not merely about the profitability of a deal at the expense of another party, but also about the circumstances in which the profit is gained—through a transaction where the other party, unable to act freely, is economically harmed.

Moreover, the disproportionate amount of profit for just one party is another characteristic of usurious practices. In fact, in some countries, conventional maximum rates are set based on the rates typically charged by market institutions. These rates are often considered fair. When an interest rate exceeds these conventional maximums, such countries deem the charge illegal, considering it usurious. However, in line with what we have explained before, the fairness of an economic value or price—such as an interest rate—is established in relation to market prices, which, as Thomas Aquinas noted, often serve as indicators of justice, but not the criterion of justice. Therefore, unusual profits according to market standards—as, for instance, interest rates much above market rates—is just an alert, not necessarily a case of usury.

4.2 An Ethical Assessment of Commercialization

A company's commercial ethics govern its relationships with commercial stakeholders—those with whom it engages in ongoing or occasional business activities. These stakeholders include customers, suppliers, and creditors. The ethical considerations in this context encompass a range of issues, from the features of the product and its price to commercial welfare and justice-related criteria. For instance, let us consider the moral legitimacy of selling beer. Any moral assessment of this everyday practice must consider not only factors such as the benefits of drinking beer, its price, its quality, and restrictions on minors, but also the broader cultural, historical, technological, legal, and even religious implications of beer commercialization.

In other words, selling beer is not morally neutral, nor bad, but not always good in ethical terms. Therefore, it is essential to have the capacity to critically assess its commercialization and, specifically, to understand the criteria necessary for the ethical evaluation of markets and products. As we will show, such principles are the criterion of common benefit, proportionality, fair pricing, moral cooperation, double effect of products and services, the importance of consumer sustainability, and the golden rule.

4.2.1 Common Benefit and Proportionality

In light with what we have explained before, the first criterion that we must consider in commercial ethics is commutative justice, that is, justice in matters of sales, usufruct, loans, deposits, and rents (Aristotle, *NE*, 1131a; Aquinas, *S.Th.*, II-II, q.61, a.1). This means that, in a fair exchange or purchase and sale between parties, a commercial relationship must be established under the principles of mutual benefit and proportionality. In fact, unfair commercial practices are what Aquinas (*S.Th.*, II-II, q.55, a.3; *S.Th.*, II-II, q.55, a.2) consider business cunning, which is the opposite of the virtue of practical wisdom or *prudentia* (Aquinas, *S.Th.*, II-II, q. 47, a.4; a.6) in economic matters in which, as long as they are virtuous, they essentially integrate both personal and common goods (Aquinas, *S.Th.*, II-II, q. 47, a.11).

For MacIntyre, "The point and purpose of monetary exchanges in markets is to enable us all to benefit from each other's labor..." (MacIntyre 2016, p. 91) and this is a matter of the common good, not exclusively of financial benefit. Hence, a new perspective on markets changes the material reason why we work or, in other words, the economic dimension of the meaning of work: "... Agents need to understand that the acquisition of money is no more than a means to the achievement of and the acquisition of goods and that such achievement and acquisition is to serve common goods" (MacIntyre 2016, p. 91).

A commitment to the common good in commercial exchanges is in fact what we can identify as the importance of being responsible for attending to the needs of others. This is, in fact, the principle of the Golden Rule applied to commercial relationships. This principle—'do not do to others what you do not want them to do to you' or, put in an active way, 'put yourself in the other person's shoes'—is not simply a matter of altruism, but a concern for others' wellbeing. This principle is a practical criterion that allows us to be impartial and benevolent spectators who can condemn certain behaviors in others, as well as forms of competition based on selfish and individualistic behavior. Furthermore, Finnis (2011a) notes, the Golden Rule allows us to see that a business or sale is unfair, even when the consumer is acting in a voluntary, legal, or informed manner, that is, knowing the potential damage that he/she can inflict if the good or service is exchanged. As such, it is a powerful determinant in moral matters as a principle that is intrinsically oriented towards promoting another person's thriving, taking into account that it only requires us to treat others the way we want to be treated. The Golden Rule, therefore, allows for personal relationships to be performed in terms of fairness, that is, according to Finnis (2011b) that each individual

> … who takes the benefits of a fair system of cooperation and mutual restraints, and service should take the corresponding burdens; and that one who has assumed the burdens is entitled to the benefits. The application of this modality of fairness is, of course, conditional on the conformity of the system with all other moral norms, including fairness in its other modalities. (Finnis 2011b)

In a similar vein of argumentations, Melé (2009) sustains that care is an essential feature of this redefinition of common good based markets; meaning by 'care' the act of trying to do your best for others, in a sense in which markets become something more than just an opportunity for individualistic satisfaction and maximization, but a responsibility to attend the needs of others through stable in profitable commercial relationships. In line with Aquinas´ understanding on a sense of common good according to which personal activities can be called as goodwill, good faith or benevolence (Aquinas, *S.Th.*, II-II, q.23, a.1; *see also* Spaemann 2000), good commercial practices for MacIntyre are not limited to proportionality in exchange (1999, p. 126), but also to a degree of care beyond individualistic economic calculation according to which a common good is served by the market.

In other words, commercial activities must be established based on the principle of mutual benefit, but not as a way for each party to achieve their own interests. On the contrary, mutual benefit must be valued because each party is committed to achieving a common good for every party involved through those same dealings.

However, the common good in commercial practices is not about equal benefits for everyone, but rather about proportional benefits. Indeed, commutative justice lies in the criterion of proportionality between both products (or services) exchanged and the deal's arrangements on pricing (Finnis 2011a). In matters of commercial justice, the exchange of goods and the sale of services are considered fair based on a variable criterion that adjusts to each party's circumstances at the time of the transaction. Therefore, proportionality cannot be fixed *ex ante*, that is, exists within an acceptable range that, under normal circumstances, fluctuates and is not determined once and for all, let alone in relation to a fixed price that ignores the decisions of the parties involved. Nevertheless, even though proportionality must consider the changing circumstances of dealings, there is a moral restraint. In ethical terms, prices—the monetary expression of proportionality For Aristotle (NE, 1133a)—proportionality in commercial exchange is defined in terms of the personal needs of the agents involved in dealings, and those needs are associated with the moral value of achieving a good living for all the community members (Aquinas, *S.Th.*, II II, q.77, a. 4; Aristotle, *Pol.*, 1252b).

4.2.2 Fair Pricing and Consumer's Sustainability

As said, commercial justice, any time it considers the good pricing derived from proportional exchange of goods, demands consideration for the personal wellbeing of the commercial parties and their common good, otherwise loses its moral status in terms of justice and common good. This, in fact, is reflected in Aristotle's theory of exchange and use value (Aristotle, *Pol.*, 1257a).

The principle of a good living, however, even though it has moral value, cannot be defined in general terms what each person needs for a good living, not even in what is minimum necessary for living with certain dignity. In fact, if society makes such an estimation, it is arbitrary, the same as price fixing by political authorities.

According to Aquinas, fair prices are usually found in what societies consider reasonable to pay for. Therefore, a just person—the wise and good man, in Aquinas words—will estimate the economic justice of a price simply by referring to market customs, that is, what the most people usually pay for when it comes to given products and services (Aquinas, *S.Th.*, II-II, q. 71, a.4). Currency, in this sense, is the common denominator that markets provide in terms of prices, that is, what is usually paid for something.

In line with this, when considering products or economic activities, the market reasonably provides a common denominator—i.e., currency—which enables a certain range of reasonable comparisons—a benchmark, we might say—to estimate priccs, costs, and profits (Finnis 2011a). The argument behind this is the fact that market actors usually ask a price for products and services in accordance with what they estimate as what they need to afford good living standards consistently with their own circumstances of living.

Therefore, referential market prices and values are not *per se* a determinant of justice but rather are 'proxies' for the criterion of fairness that each person has in economic transactions. In other words, personal judgements reflected in market prices are taken as a spectrum for fair pricing, not as a criterion for fairness. For this reason, pricing depends on the virtue of justice, particularly when market actors must take into consideration the known living standards of their counterparts, so they can make a reasonable effort to pay what is due, even when this is above the market prices (Pinto-Garay et al. 2021). In fact, if they do not, a common benefit cannot be reached strictly speaking.

On the other hand, the responsible commercialization of products and services must take into consideration the protection of consumer economic sustainability. In other words, as far as a seller is able and can responsibly do so, it must ensure that consumers with lower purchasing power do not enter the circle of poverty because of over-consumption or over-indebtedness.

From an ethics perspective, sustainability can be understood as the importance of maintaining continuous economic interaction. Commercial relationships are not just occasional; they should be long-term, providing stable income and fostering personal development. The ongoing nature of commercial activity is essential for producing goods that support employment, family life, and society. When commercial engagement is sporadic or uncertain, the ability to maintain personal and family well-being is severely compromised. For Aquinas, without the reciprocal justice of exchange, human arts and crafts would collapse, as they would lack the necessary compensation for sustainability (Aquinas, *In Eth.*, V, Lect. VIII). Thus, commercial sustainability is justified both in terms of long-term profit maximization and in ensuring the means for a good life and human flourishing.

In other words, long-term (sustainable) market relationships are a shared responsibility between commercial counterparts, particularly those with more resources and knowledge of market dynamics. Commercial sustainability, which fosters long-term relationships, provides a compelling argument for improving commerce in the interest of the public good, as it ensures a steady supply of the resources and services necessary for a good life. This steady provision highlights the need for its security, thus preventing the disruption caused by sporadic supply or the creation of insolvent market participants. As Aquinas argued, commercial activity serves not only personal gain but also plays a crucial public role by supplying the goods essential for life, particularly when providing products or services to customers (Aquinas, *S.Th.*, II-II, q. 77, a. 4).

For firms, the responsibility for consumers' sustainability is primarily reflected in the financial products they offer, such as credit cards, mortgages, loans, and refinancing services. The features of these products can significantly impact a consumer's ability to repay. When the risk of insolvency increases, it undermines the consumer's financial sustainability, leading to broader social harm. This harm often manifests as rising poverty, fueled by consumption patterns that, while beneficial in the short term for both companies and consumers, prove harmful in the long run—especially for the consumers themselves and the market as a whole. Indeed, when consumers lose financial stability, it often results in a decline in market strength, a phenomenon we might term 'commercial poverty.'

This issue, studied under the concept of household finance, becomes evident when consumers, after making poor decisions, experience heightened economic instability, illiquidity, and income insufficiency (Morduch 2023). These issues are systemic, as instability, illiquidity, and insufficiency often exacerbate one another. Accordingly, responsible selling requires active cooperation in consumers' purchasing decisions, meaning that firms must honestly assess whether it is truly in a potential customer's best interest to buy what is being offered. In some cases, this should even lead to refraining from making the sale.

4.2.3 The Ethics of Products and The Circumstances of Use

Reference to 'ethics of products' can only be sustained in a figurative sense. Just as we speak of a firm´s moral features, an ethic of products has a particular significance since neither companies nor products are responsible for themselves, but rather the people who produce, commercialize, purchase, use or consume a product or service.

Therefore, the ethics of products refers to the ethics of those who offer and sell. In this vein, when assessing the morality of selling or buying certain products or services, there are two conditions to take into consideration, commercial honesty and responsibility for the circumstances of use.

As for honesty—or the avoidance of fraud—any seller must consider three fundamental aspects, which, in the words of Aquinas (*S.Th.*, II-II, q.77, a.2), are substance, quantity, and quality. The first concerns the seller's obligation to avoid misrepresentation, such as concealing modifications made to the product to increase profit—an example being the sale of wine mixed with water. The second, quantity, pertains to ensuring accurate measurement. Lastly, quality involves ensuring that the product will perform its intended function.

As for the circumstance of use, we need to make a distinction between products and services, namely, that products are morally neutral, but not services.

Indeed, any service offered commercially is inherently a personal act of the seller. Therefore, such a service must be morally assessed as any personal act would be. This is the case, for example, with prostitution. Products, on the other hand, should be considered differently when they do not carry any intrinsic moral connotation beyond the act of selling, buying, using, or consuming them. In this regard, illegal drugs, such as cocaine, are not inherently bad; their moral implications arise only from the way they are commercialized and consumed within our societal or legal context. In other words, there are no circumstances in which the commercialization of cocaine can be morally justified. Therefore, it is ethically wrong to commercialize it. Consequently, the ethical evaluation of a product must focus on the circumstances of its commercialization. Services, by contrast, must be ethically assessed both in terms of their nature as personal acts and in relation to the circumstances of their commercialization.

Now, the commercial circumstances apply not only to the act of selling something—i.e., commercial honesty—but also, and particularly for a responsible seller, to how the sold product or service will be used or consumed. In this sense, commercial justice is not merely a matter of exchange and transactions; it also involves considering how products and services can impact the personal well-being of consumers, depending on the circumstances of their consumption and use. For example, selling whiskey to children is legally permitted in some countries, yet it is ethically unacceptable. This is due, among other reasons, to the characteristics of the consumer, which are a crucial factor in the circumstances of consumption in this case.

According to Aquinas (*S.Th.*, I-II, q.7, a.3), the circumstances of action can be classified into three kinds, namely, act, cause, and effect. The former, the circumstances of the act itself, consist in time ('when'), place ('where'), and 'the mode of acting' (what). The second kind, the circumstances of the cause of the act, are the final cause ('why'), the material cause or object ('about what'), and the efficient cause ('who'). Finally, the circumstances of effect are only those associated with the instrumental efficient cause ('by what aids'). Accordingly, our previous example, the drinking minor, is morally wrong according to the circumstances of efficient cause, that is, the 'who'.

4.2.4 Degrees of Cooperation and Double Effect

The responsibility of sellers regarding the way buyers use and consume what is sold to them—that is, in relation to the circumstances of use and consumption—is based on the principle of cooperation.

Indeed, it is reasonable to argue that buyers are generally free to purchase, use, and consume according to their own will, particularly if they are adult citizens and therefore responsible for their own actions. However, such freedom is not a sufficient argument for someone to consider it ethically permissible to sell certain products and services.

The problem in this argument is the fact that, being that consumption or use of something harmful for them or for the rest of society, the sellers have certain responsibility as they provided some needed conditions for that harmful act takes place. In the words of Aquinas (*S.Th.* II-II, q.71, a.3), is ethically wrong to cooperate with bad action "by counseling, helping, or in any way consenting, because to counsel or assist an action is, in a way, to do it."

However, there are two forms of cooperation, namely, formal and material. The first, in the words of Finnis (2011b) is associated with commanding, directing, advancing, encouraging, prescribing, approving, or defending something immoral, even though the event has not yet happened. This is, for instance, the case of social network companies that encourage the digital addiction of their customers. This form of cooperation is, in fact, co-perpetration—acting as a co-principal—when the same intention is shared. Hence, the co-perpetrator is just as guilty and responsible as the one who commits the moral wrong.

On the other hand, some sellers, without formal participation, provide what is necessary for an unethical act to take place. This is, in fact, material cooperation and, in a different sense from formal cooperation, is not always wrong. For instance, selling the car that a terrorist used in an attack is not, in the case of a seller, unethical material cooperation, unless the seller knew the purpose, that he did not share, but he did nothing about it. Material cooperation, therefore, does not refer so much to holding a similar intention to that of the wrongdoer, but to facilitating the opportunity for someone to act in a way that is ethically wrong. In other words, material cooperation refers to the responsibility that sellers might have when providing the conditions and, therefore, the opportunity and occasion for others to do wrong.

At the same time, material cooperation can have a degree of influence, that is, it can be considered mediate or immediate. The former—mediate material cooperation—refers to the provision of resources and services that, while good in themselves, allow for an ethically reprehensible action. Immediate material cooperation, on the other hand, is that which seeks to provide the necessary means for the execution of an evil action, with the intention of providing the wrongdoer with what is necessary to achieve his goal.

Moreover, material cooperation may be indispensable or dispensable. It is indispensable when the means and resources provided for another to act wrongfully could have only been provided by a single seller; otherwise, the wrongdoer would never have been able to achieve his or her goal. In a different manner, material cooperation is dispensable when the wrongdoer could have sought material cooperation from another. For Aquinas (*S.Th.* II-II, Q. 78, Art. 4, ad.3), this is the case in which " ... the usurer to whom one entrusts one's money has other means of practicing usury," therefore, there is nothing wrong in entrusting it to him that it may be in safer keeping, since this is to use a usurer for a good purpose.

To determine whether material cooperation can be considered innocent or not guilty, four criteria must be met. First, cooperative action itself must not be intrinsically wrong, except in cases of lesser evil. Second, cooperation must be motivated by a good intention. Third, material cooperation must be delayed. Finally, it must be dispensable. These criteria can be applied, for example, to ERP software providers whose clients, pharmaceutical companies, were involved in the opioid crisis in the United States.

Lastly, the commercialization of products should be evaluated in terms of their collateral impact—*i.e.*, the indirect consequences of their use and consumption in other areas of social life. A case in point is the fashion industry's contribution to fabric waste in developing countries.

Collaterals, in such cases, can be assessed under the logic of the lesser evil. According to Aquinas, such wrongdoings should be chosen to avoid greater ones. For instance, "... less harm is done by raising a false opinion in a person's mind, than by someone slaying or being slain. Therefore, a man may lawfully lie, to save another from committing murder, or another from being killed." (Aquinas, *S.Th.* II-II, q. 110, a. 2).

In this case, the ethical acceptability of selling a product that has negative collateral effects can be justified if these unwanted effects are considered lesser evil than simply avoiding commercialization altogether. This principle is known as the double effect, and it applies under certain conditions. The first condition is that the intended outcome must be of significant importance or urgency. The second condition is that the actor must aim to achieve a good outcome, while minimizing collateral damage. The third condition is that being aware of the potential collateral damage, the actor must proceed with a clear understanding of the proportionality between the intended good and the collateral effects. This requires consciously weighing obligations against potential consequences, considering indirect harm, and actively seeking alternatives to eliminate or mitigate the damage.

4.2.5 Practices Associated with Commercial Ethics

Topics in commercial matters include, among others, those related to controversial products and services, as well as commercial practices designed to defraud or deceive. Another important topic is the concept of the value chain, which, in this context, refers to a positive approach to commercial practices, emphasizing cooperation and long-term business relationships.

4.2.5.1 Ethically Illicit Products and Services

There are several products or services that are always ethically illicit because there are no circumstances of use in which the consumption of them does not cause moral personal harm than good. Among these we can particularly consider the case of pornography.

Even though legal in many countries, pornography is the only product that does not have any acceptable circumstantial use. Pornographic content, regardless of the platform, cannot be justified in any sense due to its negative social impact and its harmful contribution to the common good. It is not justified for sex education, improving intimate relationships between couples, or even boosting the GDP as a multi-billion-dollar industry. These considerations must take into account that, while education, intimacy, and economic growth are all good, the very existence and expectations set by pornography are harmful. Education is grounded in truth, yet pornography distorts the reality of sexual relationships. It depicts an unrealistic and dehumanizing form of sexual act.

Regarding its economic contribution, the pornography industry in the U.S. alone accounted for 0.5–1% of the GDP in 2019. However, this comes with a substantial social cost. Pornography leads to addiction in consumers and has detrimental effects on the physical and psychological health of actors, many of whom, after a few years in the industry, must rebuild their careers, lives, and health. Therefore, pornography is not wrong because sex is bad; rather, it is harmful because the sexual acts it portrays occur in illicit contexts, devoid of intimacy.

4.2.5.2 Ethically Restricted Products and Services

Other kinds of products or services, while not ethically illicit, have a very limited range of circumstances of use in which they can be considered morally acceptable. This is the case, for instance, with drugs or firearms.

Indeed, while there may be situations in which the use of weapons or drugs could be considered morally lawful, their production, marketing, and consumption are generally problematic. Drugs like cocaine, marijuana, or ecstasy do not have positive circumstances surrounding their production or use. The sale and consumption of these drugs support black markets, which are associated with moral evils such as cartels, human trafficking, corruption, and murder. Additionally, none of these drugs—marijuana included—has a universally proven medicinal use, and, for many, all recreational drugs have negative effects on users' health and cognitive abilities. Thus, drugs are morally illicit not because of their inherent nature but because their use entails a range of social, economic, cognitive, and psychological damages. There are no circumstances in which the production, marketing, or consumption of these drugs does not involve some negative impact on personal or social life. By contrast, this is not the case of the so-called legal drugs, whose medical benefits are proven.

Guns, by contrast, are neither inherently good nor bad. There are acceptable circumstances for their production, sale, and use, but these are highly restricted and must align with ethical guidelines. They can be used ethically when intended as a deterrent or for self-defense, though this defense must always be proportional. In other words, they are acceptable when used to prevent harm without necessarily causing death, as lethal force is not always required for self-defense.

In summary, ethically restricted products themselves are neither inherently good nor bad (with one exception), as their moral value is determined by the circumstances surrounding their production, sale, and use. Services, however, are different. Unlike material goods, services are forms of cooperation, and since they involve personal action, they are inherently moral. In general, many personal acts have little moral significance on their own—such as serving coffee in a café—but the surrounding circumstances can give these actions moral weight. For instance, if a customer suddenly suffers a cardiac arrest and, instead of helping, the barista focuses solely on serving the coffee, the act of serving coffee—though ordinarily a neutral, honest job—becomes ethically wrong.

4.2.5.3 Predatory Lending

This is a form of subprime lending characterized by discriminatory, fraudulent, and collection harassing practices implemented by some lending institutions. It refers to any type of loan targeting vulnerable populations. The business model includes high administration fees, exorbitant interest rates, and, in many cases, prepayment penalties. These loans often involve collateral and provide little information about the loan conditions. Among others, we can consider two kinds of predatory lending, as follows:

Payday lending is a predatory form of microcredit aimed at people who need immediate liquidity. Typically, these individuals have assets they can offer as collateral but do not qualify for credit from traditional banking or financial institutions. These borrowers usually have minimal financial education, and their circumstances—combined with their financial illiteracy—lead them to accept contracts they do not fully understand. The commercial practice of payday lending often focuses on aggressive sales tactics at the end of the month, when target customers find their income insufficient to cover their expenses.

Subprime mortgages, on the other hand, are loan schemes designed for individuals with low repayment capacity. As a form of predatory lending, these loans involve small amounts, but the agreements penalize prepayment, use the home as collateral, and apply very high (but legal) interest rates, typically justified by a risk calculation. This type of lending has also been packaged into debt bundles and sold to investors. Consequently, if the overall payment structure collapses—as it did in the U.S. in 2008—the investors absorb the loss, not the lending institution.

4.2.5.4 Pyramid Schemes

Pyramid schemes are a form of commercial strategy associated with multi-level marketing (MLM), which was introduced by the Nutrilite company in California in 1945. MLM, in turn, evolved from direct selling practices—sales strategies based on person-to-person marketing, as opposed to retail establishments. These schemes are also known as franchise frauds or referral chains.

In the context of person-to-person marketing, MLM schemes distribute products and services through a network of salespeople who are not formally employed or compensated by the company. Instead, these salespeople are independent associates who earn income based on their sales, but they, not the company, bear the marketing costs.

The company does not directly recruit salespeople. Rather, independent associates recruit others into the sales network, who are classified at various levels as distributors or participants (Securities and Exchange Commission (SEC) 2013). New members can further recruit additional salespeople, resulting in exponential growth potential for the marketing network.

In MLM schemes, sellers are self-employed independent distributors. Not only can they hire other sellers, but they are also required to purchase stock, and those they recruit must do the same. This structure allows the company to convert a large portion of its fixed costs into variable costs (Keep and Vander Nat 2014).

The volume of distributors grows so rapidly that only the original participants at the top of the pyramid, who provide the franchise or license to sell, make a profit. This occurs because the market is unable to accommodate the exponential increase in sellers.

Franchisors are typically aware of this and understand that the model works until those at the bottom of the pyramid realize they have lost money. These individuals often find that their earnings were limited to purchasing licenses and stock, rather than generating profits through actual sales.

4.2.5.5 Price Discrimination

Price discrimination refers to charging different amounts to different customers for the same product or service, but in an arbitrary manner. It is a form of pure discrimination in which the seller charges the maximum price to each buyer on an individual basis. More commonly, however, the seller separates buyers into groups, adjusting prices more generally according to certain characteristics.

Price discrimination can be classified into three levels. The first-level discrimination involves arbitrary price adjustments for each unit sold, even if the buyer remains the same over time. This often occurs in monopolistic markets, where the company is the sole supplier of essential products to other companies or consumers. The second-level discrimination occurs when price adjustments are made based on the volume of the purchase, charging different rates depending on how much the consumer buys. Finally, the third-level discrimination—the most common form—entails grouping customers and adjusting prices based on characteristics such as age, gender, or other factors.

Finally, it is important to recognize that price discrimination is often considered a social illness because it can lead to usury. The ability to arbitrarily set prices can exploit customers' urgent needs or financial circumstances. In this sense, price discrimination becomes discriminatory when it operates in a context similar to or akin to usury. However, if usury does not occur, a variable pricing strategy may still be considered acceptable.

4.2.5.6 Ponzi Schemes

Carlos Ponzi, an Italian immigrant who settled in Boston in 1903, devised a business scheme based on replied coupons, which were also used in the postal system. These coupons served as a prepayment method for stamps that an addressee would need to buy if they wanted their letter to be answered. Ponzi received a mailing from Spain with these coupons and realized that their value in Europe was lower than in the U.S. He saw an opportunity: if he ordered large quantities of coupons from Europe, he could resell them in the U.S. at a higher price, potentially earning a 400% profit, according to his calculations.

To fund his venture, Ponzi sought investors, promising them a 50% return on their investment. However, the scheme did not go as planned. In order to fulfill his promise, Ponzi continued to recruit new investors, using their capital to pay returns to earlier investors.

The scheme lasted for several years, making Ponzi a multimillionaire. However, when the Boston Post exposed the fraud, Ponzi was arrested, convicted, and deported to Italy, where he died penniless in 1949.

Thus, a Ponzi scheme is characterized by offering returns on investment that are well above the market rate over a relatively short period. The returns are paid to earlier investors using the capital of newer investors. Ultimately, the organizer of the scheme declares bankruptcy and appropriates a significant portion of the invested capital when the scheme can no longer attract enough new investors.

4.2.5.7 Misrepresentation

Misrepresentation refers to an assertion, verbal statement, or conduct that contradicts reality. It involves falsifying or misrepresenting (misstating) facts to obtain money, goods, or benefits of any kind that the offender is not entitled to receive. For example, a person falsely claims to represent a charity in order to receive money that he ultimately keeps; or claims that a painting is a Jackson Pollock, when in fact, it is a fake, so that he can sell it for a very high price.

An innocent misrepresentation refers to a situation in which one of the two parties, when entering into a commercial or contractual relationship, has a reasonable basis for believing that their claim is true. In this case, although ultimately false, the irrefutable ignorance present exonerates them from liability.

Fraudulent or intentional misrepresentation occurs when someone gives a statement knowing it to be false or without knowing whether or not it is actually true, causing the counterparty to purchase a product or sign a contract based on it. In this practice, there is intent to deceive.

Negligent misrepresentation refers to a situation in which the veracity of certain falsities is asserted based on carelessness and recklessness when it comes to the reality of the facts that someone in a given position should know.

4.2.5.8 Customer information

Customers must have sufficient elements based on which to make a purchase decision. It is thus important to provide information in a clear, precise, and contingent manner. In addition, it is essential to consider customers' capacity to decide, avoiding poor decisions due to excessively complex information.

In this regard, it is also important to provide information without requiring anything from the customer, in order to promote the most informed decision-making possible.

4.2.5.9 Client Suitability and Favoritism

Offering products or services must take into account customers' needs and personal circumstances, since customers are expected to truly benefit from the products and services that they purchase. In addition, it is important to act with loyalty, prudence, and care toward customers, avoiding, for example, favoritism among customers.

4.2.5.10 Value Chain Implementation

In a more strategic view of trade and market ethics, Porter (2011) and Porter and Kramer (2019) advocates for a reformulation of commercial cooperation relations. According to Porter, the capitalist system is currently facing a crisis, largely because the major market players—companies—are often blamed for social, environmental, and economic damage. The root cause of this crisis lies in corporate behavior, where the focus is mistakenly placed solely on profitability, often with little regard for long-term strategies. However, Porter argues that this should not lead to a rejection of the capitalist system itself. It is precisely this system that has the unique ability to meet needs, improve efficiency, create jobs, and generate wealth. The issue, Porter suggests, is a narrow conception of capitalism that has prevented the system from reaching its full potential in addressing society's broader challenges. To change this, the purpose of companies must be redefined: companies should create shared value, where social needs take precedence over traditional economic objectives.

Rather than focusing solely on short-term profitability, Porter envisions a company that contributes broadly to society while remaining profitable. He proposes the idea of value for improving markets qualitatively, based on three key principles:

The first dimension is the qualitative improvement of target products and markets. Porter (2011) contends that meeting societal needs in areas such as health, housing, nutrition, aging, financial security, and environmental management requires firms to critically assess whether their offerings genuinely address these domains. This includes reconsidering target markets to identify opportunities among populations excluded due to price, demand, or distribution barriers. As Kanter (2011) notes, serving underserved communities and low-income consumers can generate both social impact and business growth.

The second dimension concerns the reformulation of productivity along the value chain. Environmental and social externalities, arising from a linear model of resource extraction to waste disposal, often result when productivity is measured solely by cost reduction, economies of scale, and raw material advantages. Porter (2011) advocates transitioning toward a circular economy that prioritizes long-term sustainability. Achieving this requires coordinated principles across suppliers, subcontractors, and other actors, making it a societal effort beyond standard corporate citizenship.

The third dimension is the development of cluster strategies, emphasizing the value of geographic and conceptual proximity in fostering cooperation and a culture of value creation. Resonating with classical ethics, Porter argues that smaller, interconnected communities—comprising businesses, public institutions, academia, and social organizations—enhance mutual support and deepen engagement with non-commercial stakeholders.

These three dimensions not only shift away from the reductionist logic that equates a company's success solely with profitability, but also challenge the trickle-down economics originally proposed by Friedman, which suggested that business activities inherently benefit society.

References

Aquinas, T. (1274/1947). Summa theologica (fathers of the English Dominican Province, Trans.). New York: Benziger Brothers.

Aquinas, T. (1271/1961). Commentary on Aristotle's nicomachean ethics. New York: Dumb Ox Books.

Aristotle. 1995a. Nicomachean ethics. In *The complete works of Aristotle*, ed. J. Barnes, vol. 2. Princeton: Princeton University Press.

Aristotle. 1995b. Politics. In *The complete works of Aristotle*, ed. J. Barnes, vol. 2. Princeton: Princeton University Press.

Balot, R. 2001. Aristotle's critique of Phaleas: Justice, equality, and pleonexia. *Hermes* 129 (1): 32–44.

Beshears, J., J. J. Choi, D. Laibson, and B. C. Madrian. 2018. Behavioral household finance. *Handbook of Behavioral Economics: Applications and Foundations* 1:177–276. https://doi.org/10.1016/bs.hesbe.2018.07.004.

Finnis, J. 2011a. *Natural law and natural rights*. 2nd ed. Oxford: Oxford University Press.

Finnis, J. 2011b. *Human rights and common good: Collected essays.* Vol. 3. Oxford: Oxford University Press.
Kanter, R. M. 2011. How great companies think differently. *Harvard Business Review* 89 (11): 66–78.
Keep, W. W., and P. J. Vander Nat. 2014. Multilevel marketing and pyramid schemes in the United States: An historical analysis. *Journal of Historical Research in Marketing* 6 (2): 188–210. https://doi.org/10.1108/JHRM-01-2014-0001.
Koehn, D. 1992. Toward an ethic of exchange. *Business Ethics Quarterly* 2 (3): 341–355.
MacIntyre, A. 1981. *After virtue.* Notre Dame: University of Notre Dame Press.
MacIntyre, A. 1999. Dependent Rational Animals. London: Duckworth.
MacIntyre, A. 2016. *Ethics in the conflicts of modernity: An essay on desire, practical reasoning, and narrative.* Cambridge: Cambridge University Press.
Melé, D. 2009. Integrating personalism into virtue-based business ethics: The personalist and the common good principles. *Journal of Business Ethics* 88 (1): 227–244. https://doi.org/10.1007/s10551-009-0108-y.
Morduch, J. 2023. Rethinking poverty, household finance, and microfinance. In *Handbook of microfinance, financial inclusion and development*, 21–40. Cheltenham: Edward Elgar Publishing.
Pinto, J., I. Ferrero, and G. Scalzo. 2021. Virtuous work and organizational culture: How Aristotelian practical wisdom can humanize business. In *Human centered organizational culture, Routledge's human centered management book series*, ed. M. T. Lepeley, 118–126. London: Routledge.
Pinto-Garay, J. 2015. The concept of work in a common good theory of the firm. *Business & Professional Ethics Journal* 34:45–70. https://doi.org/10.5840/bpej20201227104.
Pinto-Garay, J., I. Ferrero, and G. Scalzo. 2021. Pricing for a common good: Beyond ethical minimalism in commercial practices. *Philosophy of Management* 20 (3): 271–291. https://doi.org/10.1007/s40926-020-00162-w.
Pinto-Garay, J., G. Scalzo, and M. Schlag. 2023. Whose work? Which markets? Rethinking work and markets in light of virtue ethics. *Business Ethics: The Environment & Responsibility* 32 (S1): 4–14. https://doi.org/10.1111/beer.12507.
Porter, M. E. 2011. Creating shared value. *Harvard Business Review* 89 (1–2): 62–77.
Porter, M. E., and M. R. Kramer. 2019. Creating shared value. In *Managing sustainable business*, ed. M. E. Porter, 323–346. Cham: Springer.
Securities and Exchange Commission (SEC). 2013. *Investor alert: Pyramid schemes posing as multi-level marketing programs.* Washington: SEC.
Spaemann, R. 2000. *Glück und Wohlwollen: Versuch über Ethik.* Stuttgart: Klett-Cotta.

Chapter 5
Corporate Ethics

Abstract This chapter explores corporate ethics as the set of responsibilities that firms bear toward the public good, understood through the framework of legal-political justice offered by New Natural Law. Business and society are not separate spheres but intrinsically interconnected: firms depend on social order, taxation, and regulation, while their activities directly shape public life. Corporate ethics is therefore framed as civic responsibility, where firms contribute to the common good of the polis by sustaining legality, paying taxes, and respecting institutions. The chapter examines the role of the state as the guarantor of social peace, clarifying the moral obligations of companies regarding taxation, duties, and the rejection of avoidance, evasion, and bribery. Beyond compliance, firms are called to cultivate civic virtue by supporting intermediate associations—such as cities, schools, and cultural institutions—and by fostering corporate citizenship through urban development, philanthropy, and values that transcend profit maximization. Finally, the chapter highlights the social dimension of work and the importance of unions as mediators of justice, as well as environmental stewardship and sustainability as extensions of the firm's responsibility to future generations. Corporate ethics thus integrates economic success with civic participation, justice, and care for the common good.

Keywords Corporate ethics · Legal-political justice · Common good · Corporate citizenship · Taxation and duties · Bribery · Civic virtue · Unions · Sustainability · Environmental responsibility

The relationship between business and society—understood as corporate ethics—is not grounded in the interaction between two separate or adjacent spheres, as implied by the usual distinction between the private and public realms. Rather, business and society are intrinsically interconnected, since employment and markets are essential components of society, just as society is foundational to their functioning. However, the concept of society lacks the specificity required to define corporate ethics through the Aristotelian lens of justice. Accordingly, the notion of 'public' instead of 'social' proves more precise, as it better captures the responsibilities that firms hold toward

J. Pinto-Garay and G. Scalzo, *Justice and Corporate Excellence*, Ethical Economy 72,
https://doi.org/10.1007/978-3-032-12189-9_5

third parties beyond their organizational and commercial relationships (Scalzo et al. 2021).

Business and society intersect in several important dimensions of public life, such as tax policies, the regulation of monopolies and collusion, the creation of labor laws and courts for commercial justice, urban planning that accommodates growing production and commercialization, and the protection of private property, among others.

In this context, certain individuals, groups, or institutions connected to a firm's activities should not be considered commercial or organizational stakeholders, but rather public stakeholders. This distinction arises because the justice relationship between these entities and the firm is based on legal compliance and a broader responsibility to the public good, specifically the common good of the city (*polis*) (Sison and Fontrodona 2012, 2013).

In this context of business-society relationship, the principle of responsibility is grounded in Aristotle's (1995, 1095a) conception of humans as political animals, naturally inclined to participate in various communities, particularly the city (*polis*). Being a political animal, in Aristotelian terms, means that every person seeks a good life within the city, which is only possible within a public community characterized not just by lawmaking (Dierksmeier and Celano 2012) and compliance, but also by active participation. This participation involves integrating both legal and extra-legal norms—social and moral principles—governing fair relations between groups, institutions, communities, and political authorities, i.e., civil society (Finnis 2011b). Legality and regulation represent the basic standard of civility. As Finnis (1998) further explains, the 'public' is defined by social order and participation, rather than merit or mutual benefit, as seen in distributive or commutative justice.

For this reason, legal-political justice is sometimes referred to as social justice, reflecting the need for every member of the community to cooperate in creating and maintaining the institutions and forms of life that are preconditions for long-term sustainable flourishing within the city. Social or legal-political justice is, therefore, a matter of citizenship and public life (Finnis 2011a). The common good of social justice, as Finnis describes it, concerns the physical resources and structures of expectations and dependencies essential to the well-being of all community members, especially the weak and vulnerable (Finnis 2011a).

> Similarly, public order, in its restricted, common law sense, concerns the maintenance, not so much of the psychological substratum for mutual respect, but of the physical environment and structure of expectations and reliances essential to the wellbeing of all members of a community, especially the weak. Inciting hatred amongst sections of the community is not merely an injury to the rights of those hated; it threatens everyone in the community with a future of violence and of other violations of right, and this threat is itself an injury to the common good and is reasonably referred to as a violation of public order. Rioting and bombing, and threats thereof, are not merely prejudicial to the rights of those killed or injured, but to everyone who has now to live in a community where such things happen. (Finnis 2011a, p. 217)

Thus, corporate activities aligned with the public good are fundamentally about supporting social peace—something that can be undermined by actions, attitudes, or omissions that harm public society. This support is bounded by the firm's essential responsibility for its own areas of competence, namely employment and commercialization. In this sense, Finnis observes that practices based on selfish individualism, contentiousness, obstinacy, quarrelsomeness, or sedition against authority—contrary to participation and integration in public harmony—undermine the social order and are contrary to the values of justice and citizenship (Finnis 2011b).

5.1 The State

From a social justice approach, the relationship of any firm with society inherently involves legality and political authorities. In its political-legal dimension, social justice concerns political authorities (ST II–II q.58 a.6c.) and, by extension, the broader community, at least, in its role as citizens committed to both the function of the state and the obedience of its laws (Finnis 2011a).

Regarding the former, the modern state was initially conceived not only as an alternative to monarchy but also as a distinct sphere from civil society in terms of its functions, although it relies on civil society for resources. In effect, the modern state is an administrative body sustained by taxation, leaving productive tasks to the market economy. Its role, therefore, is solely focused on ensuring the general conditions for production through a legal framework and the necessary infrastructure for the workforce (Habermas 1999), upon which, it is important to note, it depends economically. Therefore, the state is legally a reality that possesses both internal sovereignty—governing itself—and external sovereignty, extending over the nation and its territory. In its social sphere, the state has sovereignty over all its members—its citizens—and its power is defined through positive law. Additionally, the state is constituted in relation to a nation, which gives it the connotation of a political community defined by shared language, culture, genealogy, and history.

However, the state is a political institution, not a political community. It serves as the administrative organ of a political community, at least in the classical sense that distinguishes the community from intermediate bodies and the family (Finnis 1998). Thus, the state is not synonymous with civil society, as citizens do not necessarily shape the state's activity, though they are intrinsically related to it. Nor is the state equivalent to the nation, though the nation—which includes cultural, linguistic, historical, or genealogical elements—is defined in relation to a territory that the state is responsible for safeguarding.

Similarly, although the state is distinct from the market, it plays a crucial role in enabling the functioning of markets and industry by maintaining order and ensuring social stability (González & Scalzo, 2024). While a separation between the state and the market does exist, as previously discussed, their relationship is shaped, on the one hand, by the tax obligations of companies, and on the other, by the state's responsibility to provide the necessary conditions for labor and market activity.

As John Finnis (1998) explains, the role of the state's administration—or public authority—is to pursue the public good in accordance with an aspirational conception of the common good. In this regard, Aiyar (2000) explains that, according to Finnis, human flourishing consists in the realization of basic goods and society, therefore, must be ordered in such a way that this flourishing and the common good of the community are made possible. This ordination by the public authority, in fact, achieves the conditions for the community members to flourish by safeguarding social peace. Therefore, the administration is duty-bound to restrain actions and words that undermine peace, concord (the tranquility of order), and the availability of resources necessary for living well. In this context, paying taxes to cover the administrative costs of the state is a responsibility to the social common good In line with Aquinas, Finnis explains that civil government proper function is the provision of peace, which means (1) absence of words and deeds immorally opposed to peace. (2) Concord (i.e., tranquility of order) and perhaps (3) a sufficiency of at least the basic necessities of life (Meredith 2008).

Therefore, the moral obligation to obey the law is not based on a preeminence of the state over society, but on the need to safeguard social order through formal agreements between parties. As Finnis explains, unlike promises between citizens, which can be ambiguous, the law is far more effective as a means of maintaining order, rooted in a responsibility for social justice and the common good, which is objectively defined by society through its authorities—without relying on individual convenience or a biased conception of the social good (Finnis 2011a). In this sense, the moral obligation to obey the law is owed to fellow citizens rather than the political authority represented by the public institution of the state (Finnis 2011a).

In this context, it becomes clear that businesspeople and firms have a responsibility to the public good not only by complying with the legal order, but also by contributing to public authorities and institutions through the payment of taxes and duties.

In general terms, taxes—from the Latin *tributum*—refer to payments made by individuals and institutions based on their income to support public administration. Taxes contribute to the public good by funding state expenditures, often in a manner proportional to income (i.e., those who earn more, pay more) or through specific actions, such as tolls or VAT payments. Ultimately, taxes are a contribution toward the public good, which depends on the efficient functioning of the state rather than the state's functioning itself.

According to Meredith (2008), Aquinas maintains that taxes are a stipend to public authorities, that is, for their maintenance, not a reward for their service, but the living to which they are entitled as ministers in return for the preservation of public peace. By contrast, duties (*vectigal*) are payment to public authorities for other public services different from those associated with public peace, like keeping highways and roads.

Accordingly, those engaged in commercial activities—whether directly or through institutional means (i.e., firms)—can be required to pay taxes and duties as a means of making their fair contribution to the public common good, which is, as explained before, social peace.

5.1.1 *Practices Associated with Business And State Relationships*

In view of the preceding discussion concerning the role of the State—particularly with respect to the concepts of taxation and civic duties—it is pertinent to examine certain related practices. Furthermore, it is necessary to address at least one practice directly associated with public authorities: bribery.

5.1.1.1 Tax avoidance, Evasion, and Tax Exemption

The normative scope of tax payment may be partial in relation to the complexity and volume of commercial activities subject to taxation. Tax regulations, in fact, do not always anticipate all activities that should be taxed, which can result in regulatory or legal loopholes. These loopholes allow, but do not explicitly authorize, the non-payment of certain taxes.

Taking advantage of these loopholes, however, goes against the spirit of the law, even if it does not involve blatant offenses. Tax havens, for example, foster avoidance practices that, while not illegal, are contrary to the public good and the intent of the law. The same is true when a taxpayer increases expenses to reduce taxable income.

Tax avoidance is ethically illicit, even though it may not be illegal. It is not strictly legal either, as it does not incur penalties or sanctions because the law has not yet addressed specific business practices that should be taxed. Avoidance is illicit because it undermines the collective good of the community from which the taxpayer benefits.

On the other hand, tax evasion is clearly illicit because it is both illegal and immoral. It breaches the duty of justice owed to fellow citizens, who, by paying taxes, support the functioning of the state. Economic activities—whether individual or entrepreneurial—are made possible by a functioning state, and all economic actors have a duty to contribute.

When other citizens contribute to the state through taxes, they are indirectly supporting their own economic activity. Tax evasion, therefore, constitutes the usurpation of economic resources from others, akin to theft or misappropriation. Furthermore, tax evasion demonstrates a disregard for the common good, as tax payments are vital to providing public benefits that society needs to thrive. The payment of taxes is a duty of justice to other members of society, and failing to pay taxes reflects a form of individualism that undermines the fundamental principles of community life.

It is important to note that paying taxes does not directly correspond to the payment for specific public services, such as police, public parks, or justice. Public services cannot prioritize those who pay the most taxes; rather, they must address needs that are urgent for maintaining social peace, often affecting those who pay the least. Therefore, the payment of taxes is an act of justice and solidarity within the community, as all citizens contribute to safeguarding the collective good and social peace.

Paying taxes is not merely justified by the provision of specific public services, but also by the long-term preservation of the institutions that provide those services, ensuring the continued order of the state.

When taxes are directed toward the public good—rather than the benefit of the state—it is understandable that some legitimately owed taxes may be excused in particular cases. This practice, known as tax exemption or waiver, involves suspending a tax obligation (even if the taxable event has occurred), forgiving a debt (even if the obligation is recognized), or refunding taxes already collected. Authorities may forgive debts to reduce the risk of bankruptcy and preserve jobs, which are vital public goods. In these cases, the fundamental social principle of tax payment is not suspended, but the requirement is temporarily relaxed for the greater public good, given the taxpayer's circumstances, whether a business or individual.

Exemption, in whatever form (suspension, remission, or reimbursement), is a circumstantial act that cannot be applied as a general rule and must be exercised prudently, with consideration for public well-being. In addition to such circumstantial cases, companies in various countries may receive permanent tax benefits based on their contribution to the public good. For example, the state may suspend VAT obligations for companies delivering essential products and services in remote areas.

5.1.1.2 Unjust Taxing

At the same time, we must consider the problem of unjust taxation—specifically, whether it is morally acceptable to contribute taxes to the functioning of a corrupt, illegal, inefficient, genocidal, totalitarian, or otherwise unjust state apparatus. The refusal to pay taxes in such cases is known as civil disobedience. To address this issue, we must generally consider the principle of moral cooperation and the criteria for determining when a state is unjust.

As explained in business ethics, moral cooperation refers to indirect participation in an act performed by a third party. This secondary participation, while not direct, still makes us responsible to some degree. The most common form of moral cooperation is material cooperation, which involves providing material resources that enable the commission of an intrinsically wrongful act by another person, group, or institution. In order for material cooperation to be culpable, it must also be formal, meaning the person providing the resources must know that they will be used for illicit purposes by a third party. For example, paying taxes to a genocidal state constitutes illicit moral cooperation if the taxpayer knows that the funds will be used for such purposes.

However, we must also consider the direct effectiveness of the resources provided. For instance, in the case of a genocidal state, if the tax paid by a single individual is relatively marginal compared to the total national contributions, withholding payment would not significantly impact the state's ability to carry out its atrocities. Therefore, the decision to pay or withhold taxes may have little effect on the state's actions, and refusing to pay becomes more of an act of personal conscience.

Nevertheless, formal cooperation remains in place, and moral cooperation continues. Thus, an argument for civil disobedience still holds, but it requires a nuanced analysis. A taxpayer's circumstances may mean that withholding payment could result in greater harm to others—such as if the taxpayer's imprisonment leads to the suffering of their dependents.

Therefore, civil disobedience in the context of taxation—even when the taxpayer is aware of the state's illicit actions—must be evaluated in light of the individual's personal responsibilities and the potential harm their actions may cause to others.

5.1.1.3 Under and Overtaxing

Another form of unfair taxation can be seen in what is called an exaggeratedly high tax rate (overtaxing), which hinders the normal development of a taxpayer's daily life or general economic progress. This occurs when a high tax burden prevents individuals from fulfilling their essential responsibilities, such as parents struggling to support their children's maintenance, education, health, and well-being.

However, categorizing a tax policy as overtaxing is often difficult in terms of objective and accurate judgment, especially in the two areas where this discussion typically arises: inefficient public administration and/or public policies aimed at reducing social gaps in the short term.

Regarding inefficiencies in public administration, overtaxing is seen as such when it is clear that the state is overspending, making untimely expenditures, allocating resources to initiatives outside its scope of responsibility, or engaging in outright corruption.

Yet, even in these cases, it is important to recognize that citizens are not entirely objective when assessing public management. Perceptions of overtaxing may be influenced by subjective judgment, as efficiency in public spending is often compared to a hypothetical scenario of perfectly effective administration—one that assumes the existence of an objectively prioritized hierarchy of spending perfectly aligned with the balance between citizens and the state. Thus, it is difficult to claim that citizens always accurately assess public spending or that a tax policy is truly overtaxing simply because they perceive it as such.

The second issue—addressing inequality—is undoubtedly more complex. Inequality undermines social peace, which political authorities are responsible for promoting and maintaining, and theoretically, higher taxes on wealthier individuals could help reduce social gaps.

Piketty's (2014) proposal is the most prominent in this regard. The French economist distinguishes between the concentration of capital (wealth) and income. For Piketty, income is recurrent and corresponds to the quantity of goods and services produced and distributed annually. It can be derived from labor (salaries, fees, bonuses, etc.) or capital (rents, dividends, royalties, etc.). Wealth, on the other hand, represents a stock and reflects the total wealth available at a specific point in time (private wealth). This wealth results from acquisition or accumulation over time and can be considered non-human net worth.

According to Piketty, inequality in economies like those of Europe or North America is more severe in terms of wealth than income. While income inequality exists, it is less extreme and less controversial when tied to merit. The solution to inequality, therefore, lies in taxing wealth, not income. This would include taxing inheritance, assets, and so on. As for income inequality, Piketty attributes part of the issue to the so-called 'skills race' driven by globalization and technological gaps in job preparation. However, Piketty argues that this explanation is limited, as Europe and Japan, despite similar educational and technological advancements, have narrower inequality gaps than the U.S. Hence, the underlying problem seems to be more closely related to wealth (from a tax perspective) than income or labor.

If Piketty is correct, then a tax burden aimed at reducing these gaps would not constitute overtaxing, but rather sound public policy. However, while social instability is certainly detrimental to the common good, a wealth surtax alone does not guarantee that the funds collected will be effectively redistributed by the state. Therefore, Piketty's proposal could still be seen as a form of overtaxing if there is no clarity on how the revenue from taxing the wealthy will be utilized.

5.1.1.4 Bribery

In relations between business and the state, one type of malpractice is bribery or subornation, which is also referred to as extortion, illegal exaction, perversion of justice, kickbacks, or venality. These terms all describe acts through which a private entity, such as a company, influences a public or state authority to act in its favor in an exceptional and illegal manner, contrary to both regulations and the duties of the public office held by that authority.

This form of corruption is ethically illegal for at least three reasons: (1) it undermines the function of the state, (2) it negatively impacts the market, and (3) it is unjust to the rest of the citizens.

Regarding the first reason, bribery involves payments to a public official or authority to perform, favor, or prevent the performance of certain duties that fall within their job description. Such conduct is improper not only because it supports the private interests of an individual rather than the official's professional duty but also because it prioritizes personal economic or other interests over the official's role and the institution they represent. This ultimately shifts focus from the public good to private interests, which may or may not conflict with the law.

On the second point, bribery undermines the market by granting commercial, operational, or economic advantages to those who engage in illegal practices, thereby violating the principle of free competition. Free competition encourages improvements in production and commercial processes rather than granting unfair advantages through public influence.

Finally, the perversion of justice occurs when favoritism is shown toward some citizens at the expense of others. Such illegitimate privileges violate the principle of equality and justice, which mandates that authorities serve citizens' needs based on the public good, general interest, and the law, rather than according to the financial contributions of those seeking preferential treatment.

Moreover, acts of corruption are not limited to businesses; they can also arise from government authorities who demand illegitimate payments for businesses to operate legally. This form of malpractice, commonly referred to as bribery, can be particularly problematic for companies operating internationally.

In cases where bribery is a culturally accepted or habitual practice in a foreign country, the general principle is to avoid engaging in bribery and, if possible, suspend business activities there. However, the principle of lesser evil may apply, especially in situations where the company's operations provide a livelihood for local employees who depend on these activities for their subsistence.

5.2 Corporate Citizenship: A Matter of Civility

In the context of the business-society relationship, a firm also plays a public role, with a responsibility to contribute to the public good through its civic virtue. This responsibility involves associations with institutions other than the state and the family. In this sense, firms establish relationships with what are referred to as 'intermediate associations'—including schools, cities, shops, marketplaces, highways, churches, charities, sports organizations, and other voluntary associations, as well as professional dealings and state functions such as jury trials, elections, military service, public administration, and legislative activities (Finnis 2011b).

Civic virtue, in this context, is defined as excellence in the civic dimension of a political community—specifically in the relationship between fellow citizens (Lovett 2015). Excellence, as Aquinas describes it (*S.Th.*, II-II, Q. 26, Art. 3), refers to the habit of prioritizing the well-being of fellow citizens over one's own material benefit, especially in situations where a community member is in need (Finnis 2011b).

An essential aspect of civility is the respect and appreciation of diversity—much like a climber values the mountain—seeing differences as opportunities for overcoming obstacles and fostering mutual assistance. Despite the natural differences among members of any civic community, civility involves respect for humanity and the dignity of fellow citizens, particularly those in need. In this way, civility contributes to a richer and more authentic human fulfillment for all (Finnis 2011b).

Since virtues cannot be considered as inherent to institutions—only individuals within an institution can be regarded as virtuous—the civic character of a firm, in terms of excellence, must be understood differently. An institution itself cannot possess virtues, but it can reflect the moral qualities of its members, particularly those who hold leadership roles and make decisions that guide the firm's actions. In this sense, when members of a firm personally embrace the importance of acting with civility and based on this principle, lead corporate initiatives accordingly, we can say that, by analogy, the firm has the capacity to demonstrate excellence in its civic relationships. Thus, outstanding firms in terms of their civic character are those that have addressed the needs of other institutions or individuals in civil society, even beyond their legitimate interests in productivity, commercialization, and profit.

5.2.1 *Practices Associated with Corporate Citizenship*

As follows, we will explore several important topics derived from the way firms engage with society in terms of citizenship—specifically their role in contributing to the well-being of other community members and institutions beyond their productive and commercial activities.

5.2.1.1 Urbanity

It is difficult to understand the existence of modern business outside the context of cities. Since the eighteenth century, businesses, whether corporations or closed enterprises, have existed as a function of a lifestyle that is specific to the city. As Sombart explains in *Luxury and Capitalism* (1967) and as we have already mentioned before– the first author of the theory of entrepreneurship was Richard Cantillon; of note, this economist was the first, in 1755, to use the word entrepreneur to refer to commercial institutions, in the context of trade as an essentially urban activity.

For Sombart, modern capitalism is largely due to the emergence of cities or towns. In a synergistic relationship, cities historically expanded because they are home to large swaths of consumers who concentrate consumption in one place and stimulate the growth of the capitalist system. The consumption of so-called luxury goods (i.e., ones that go beyond what is necessary), rather than mere commerce or industry, makes this expansion possible.

In this sense, modern enterprise is a historical and sociological reality that cannot truly be understood in terms of its activities -and particularly in terms of an ethical assessment of its functions- outside the geographical and cultural framework of the city.

For Sombart, in fact, the importance of the city in the development of capitalism lies in having offered a joyful and exuberant lifestyle—as new forms of social luxury—that included not only the upper classes, but all the inhabitants of the city. It is a passage from private luxury to collective luxury. Thus, the city offers theaters, dance halls, restaurants, hotels, stores, and museums, that is, new forms of business development that are by nature, urban.

Thus, modern business is an essentially civic institution, meaning it cannot be understood in its origin or function apart from urban activity and development. Therefore, corporate citizenship does not refer to the articulation of two distinct and opposing spheres (such as private and public) but rather emphasizes a dimension that is inherent to corporate activity, namely, its urban nature.

In this regard, it is worthwhile to ask how companies can facilitate a qualitatively superior lifestyle for their fellow citizens or, in other words, how companies can contribute to a better city or, by analogy, be better citizens. In our view, they can do so by integrating their commercial activity more strongly through three areas that are proper to the city, such as urbanism, architecture, and urban design; distance and mobility in the city, and access to a city lifestyle.

As Guillén (2006) explains, the idea and use of modern cities and buildings-modern architecture-is due to the relevance of modern industry and its role in the creation of new materials and work styles that impact the way people live in cities. In this sense, titans of modern architecture, such as the Bauhaus movement or Le Corbusier, created a new way of thinking about industrial and public spaces based on a close relationship with industry and management theory, and in particular, the scientific and rationalist principles of Taylor and Ford.

Thus, the Bauhaus can be understood as the most important intellectual movement in modern industrial design. Founded in 1919 by architect Walter Gropius (who was convinced of the value of standardized production and scientific management), the movement substantially contributed to the modern design of furniture, lamps, everyday items, wallpaper, graphic design, prefabricated houses, and buildings. Gropius' work was notable for its close collaboration with industry and the development of real estate projects for workers.

Similarly, the development of modern products, buildings, and cities is due to the architect Le Corbusier (AKA Charles Édouard Jeanneret), who integrated the principles of twentieth century industrialization into the idea of architecture. Indeed, for Le Corbusier integrating the rationality of business administration into architecture was essential so that the latter would lend rational solutions to the use of cities and things. Following these ideas, several French architects designed stores, industries, apartments, offices, buildings, and houses, in addition to the development of urban design principles known as 'Municipal Taylorism' (Guillén 2006). Therefore, architectural and urbanistic innovation became evident along with a new spatial nature of companies and, vice versa, the industrial and corporate sense of architectural thought. Companies are, in this sense, by nature, a configurator of cities, in the same way that cities configure business activity.

Thus, the modern city is, to a large extent, the result of how the company understands itself, its idea of work, production, the market, and consumption. Moreover, cities develop in step with new technologies and products offered by companies. The ancient social division of labor that Aristotle speaks of, and according to which the social life of the *polis* is ordered, is now a social division in which companies participate; in many ways, the contemporary city develops in terms of this participation. Thus, corporate citizenship is not just a strategic relationship with social actors; rather, it represents a qualitative improvement to the citizen nature of companies that makes the city a better place to live.

This, as mentioned, materialized in various aspects that shape urban life, such as commuting policies for workers and the creation of private spaces for public use (parks, shopping malls, terraces, etc.). It also includes the aesthetics of the city through the edification of corporate buildings, real estate development that contributes in economic as well as cultural terms, the use of efficient materials to save energy and those same materials' contribution to silence (avoiding noisy or acoustically polluted cities) and safety, the sponsorship of public activities, sports, social integration through common spaces for recreation or culture (e.g., museums), etc. Corporate action in the city thus assumes a leading role in the creation and use of cities.

5.2.1.2 Philanthropy

Since the publication of Milton Friedman's article on the purpose of business and its social responsibility (*The Social Responsibility of Business is to Increase its Profits*), business activity has often been understood by various authors as purely economic, with everything else considered philanthropy. This view, known as the Separation Thesis (Alzola 2011), asserts that all non-economic activity lies outside the scope of business. Exceptions to this thesis typically refer to the commercial value of non-economic social activities, such as marketing or corporate reputation, or the potential for tax breaks.

However, beyond the theory of convergence between philanthropy and commercial activity, companies often engage in philanthropic endeavors that are integrated with their business activities. These initiatives aim to address social needs as part of a broader commitment to contributing to the community. In this context, corporate ethics should emphasize the altruistic nature of corporate activity rather than its strategic advantages.

The etymology of 'philanthropy' itself reflects the essence of the activity: derived from the Greek *philos* (friendship) and *anthropos* (human), it refers to acts motivated by love for humanity—an active concern that seeks to overcome poverty and indignity. Philanthropy, therefore, is justified by the presence of poverty, understood broadly, not just in material terms.

Corporate philanthropy, then, seeks to address both circumstantial and structural poverty outside the company. This can take the form of donations, campaigns, corporate foundations, and partnerships with nonprofit organizations. The ultimate goal of these efforts is to help individuals who experience poverty in their various forms—circumstantial or permanent, objective or subjective, relative or absolute.

Circumstantial poverty refers to those who, despite having previously enjoyed a good life, find themselves in difficult circumstances due to rising living costs, health issues, natural disasters, economic crises, and the like. This aid differs from efforts aimed at alleviating structural or permanent poverty, which affects individuals who have always lived in such conditions. Corporate philanthropy, therefore, may provide both immediate relief for those in temporarily need and long-term support to address systemic social poverty.

While material poverty requires urgent attention, aid for subjective or cultural poverty is also significant. This is because one of the major causes of material poverty is the lack of cultural capital, education, social networks, friendships, or family connections.

Poverty can be understood in two ways: relative and absolute. Relative poverty refers to the disparity between those who have enough—or more than enough—and those who have little or nothing. Absolute poverty, on the other hand, pertains to a lack that directly affects the humanity of those in that condition, regardless of comparisons with others.

Both forms of poverty should inspire philanthropic action, translating into solidarity and efforts to restore dignity. In the relative sense, poverty may not be an issue of social justice—i.e., the equilibrium of society—but of solidarity, driven by the desire for all individuals to have access to material and subjective goods enjoyed by others.

In the absolute sense, philanthropy addresses the indignity of those living in poverty. However, it is important to be cautious when justifying philanthropy purely as a matter of social justice, as this approach can shift the focus onto those who have more, rather than recognizing our own privileges or, more importantly, the humanity of those in poverty, who risk being dehumanized

5.2.1.3 Corporate Axiology

Axiology, derived from the Greek *axios*, meaning 'value,' refers to something estimable or valuable. It is not a moral principle or a virtue but rather a personal, community, social, or political goal that can influence lifestyles and institutional practices. Values of this type, such as political, social, or lifestyle values, reflect ideals—visions of a way of life to be achieved or preserved.

The theory of value explores the relationship between the ideal configuration of the human world, the psychological aspects of valuation (what seems satisfactory to us, whether it is anticipated or already achieved), and its moral nature—considering its real moral goodness. A value (such as equality, health, justice, wealth, etc.) inherently has a moral nature. However, the fact that something is considered a value does not necessarily mean it seeks something good. For example, the Nazis held values that, although they viewed them as good, were objectively harmful and unjust.

A theory of corporate value is crucial not only conceptually but also practically, as companies—especially large corporations—are driven by their values. Reducing companies to institutions focused solely on profit maximization limits their 'values' to a narrow sphere, namely, the value of profitability.

Corporate life, however, is sustained by a set of values to which profitability is subordinated. For example, the definition of a company's mission or vision is an important axiological-corporate act that reflects how profitability and economic success are understood. A company's values, which define its purpose, are ideological forms with significant content, whether positive or negative. These values fulfill an ordering function and provide the company with social and commercial identity, as well as legitimacy in the eyes of stakeholders (Scalzo et al. 2020). Corporate axiology, in this sense, explains the company's value structures, which are oriented toward functional, reputational, cultural, and political aspects.

Values are functional to an organization when they enable strategic decisions, team-building, or commercial presence. For instance, the value of sports for Nike or Puma, safety for Volvo, or speed of service for Burger King—all represent functional axiological pillars within the company, regardless of their theoretical origin. The idea of health, service, or access to environmentally friendly products may stem from a founder's ideals, a company's strategic reformulation, or social demand. However, the origin of these corporate values is secondary to their role in guiding company practices.

From a social perspective, however, the origin of corporate values is highly significant. Social values embedded in a company, such as environmental responsibility at Shell or Exxon, are more or less valuable and credible based on the sincerity of their commitment to the value (e.g., real care for the environment) rather than instrumental reasons (such as brand prestige and commercial performance). Similarly, corporate values such as gender equality, inclusion, and work-family balance gain social credibility when they are genuinely valued for their own sake, rather than merely for their potential economic benefits or reputation.

In addition to functional and social corporate values, corporations can also adopt political or ethical values aimed at implementing certain principles within the company and society. For example, companies like Fox News or CNN have explicit political agendas—Republican and Democrat, respectively—Huawei aligns with a communist agenda, and the fast-food chain Chick-Fil-A embraces evangelical Christian values. In these cases, the formulation of values is more aligned with internal or social activism by the company rather than functional objectives like those seen in Nike's emphasis on sports. Therefore, the dissemination of values beyond the corporate sphere should be approached with caution, as such actions can blur the company's role, which is distinct from that of a political party, NGOs, or think tanks.

5.3 Unions: A Social Dimension of Work

According to the scheme of justice, the most important social aspect of fairness in labor is the moral obligation of employers to provide dignifying working conditions and to pay their employees with adequate compensation for workers to have suitable living conditions.

In this regard, a social or legal-political criteria of justice has not only led to regulations on labor markets—*i.e.*, minimum wages or working conditions policies—but also to the protection of workers associations, that is, unions or guilds. Accordingly, an essential aspect of this social dimension of work according to a social justice approach is the right of workers to associate in order to improve their employment conditions. Thus, the right to associate safeguards the formation of associations or unions aimed at defending the vital interests of individuals who, in this case, work in various professions and occupations that make up a society (Wojtyla 1981).

5.3.1 Rights and Bargaining

The history of modern trade unions is closely linked to the expansion of labor rights. The Industrial Revolution of the eighteenth century, which brought about unjust working conditions—such as sweatshops—highlighted the need for bargaining power to improve both working conditions and wages. In this context, the issue of labor rights developed through legally supported unions aimed at balancing the power dynamics between workers and employers. In other words, the social and legal protection of unions facilitated a more horizontal relationship between workers and employers, fostering fair employment within the framework of distributive-contributive justice. In this regard, the role of workers' organizations in modern times became quite different from the old guilds.

Indeed, the concept of a union, in its etymological origin, derived from the Ancient Greek word '*syndikos*' (συνδικος), which refers to someone responsible for administering justice (δικειν) through protective functions. In other words, a syndic is a protector in matters of rights and justice. Later, in Latin, the term 'syndicus' was used to refer to someone who fulfills the role of a defender. By the sixteenth century, the term came to denote civil magistrates or leaders with a representative role for groups. Unions, therefore, are characterized by their protection and defense of rights, whether in relation to employers, authorities, or industrial sectors.

As for guilds, the workers' association serves a different function. From Latin *gremium*, this concept originally refers to a lap or bosom, that is, the central place where the parts of the body are articulated or assembled. *Gremium*, in turn, derives from *gregex*, which can be translated to Spanish as *grey*. From here, the verb *grego* means to add or assemble. In its etymological sense, *gremio* refers to the congregation of people in an organized or assembled body.

Historically, modern guilds trace their origins to the middle ages. As Wojtyla (1981) explains, they emerged from medieval artisan corporations that brought together individuals from the same profession or those performing similar types of work. These artisans were initially independent workers, such as luthiers, builders, and goldsmiths. The original purpose of these guilds was broad, primarily focused on protecting common interests rather than representing workers before employers. These interests included, among other things, training apprentices, controlling prices and the supply of products and services, and fostering fraternity initiatives to support members in need or difficulty.

In a different manner, modern unions—in part inheritors of the guild tradition—not only serve the function of defending the rights of contracted workers but also fulfill a role that materializes in social assistance tasks for their members. However, what is a particular and essential feature of unions is the defense of rights, not the provision of assistance to their associates.

Now, the idea of the right of workers to associate in unions or guilds from a neo-Aristotelian understanding of justice is not the same as what nowadays is defended as the right to work—in fact, the most accepted idea of social justice is far from the neo-Aristotelian definition that we have provided here.

The modern concept of the right to work was introduced and popularized by the French socialist historian and politician Louis Blanc in the mid-nineteenth century. A century later, built on Blanc's vision, the United Nations General Assembly incorporated this right into the Universal Declaration of Human Rights. The declaration affirms that the right to work includes the freedom to choose employment, the right to work under fair and satisfactory conditions, and the social right to seek protection against unemployment—whether from the authorities or, in certain circumstances, from employers, through the right not to be dismissed.

In line with Blanc's definition, union activity is clearly associated with representing workers to employers to ensure suitable working conditions and to authorities in matters of unemployment. The principle of freedom of choice does not seem to be one that unions can effectively protect in practice.

In addition to Blanc's idea of workers' rights, Marxism added a second characteristic to unions activity, that is, the struggle, protests, and strikes, not as a way to reclaim fair conditions of work, but as a natural way of acting for workers associations. Indeed, the Marxists ideology attempted to explain the nature of work according to the idea of class struggle and the need to end the supposed capitalist system that alienates workers whenever production is subject to the division of labor.

From a neo-Aristotelian perspective on justice, the idea initiated by Karl Marx is, in fact, unjust. It represents a political project aimed at reshaping the social order in favor of a single group—the proletariat or working class—at the expense of others, namely the capitalists. Moreover, class struggle as an explanation of social dynamics is fundamentally opposed to collaborative systems and, by extension, to the principle of the common good.

In this sense, it's important to recognize that union activity does not necessarily align with Marxist ideology. On the contrary, a union—even one with a more politically left-leaning stance—has the responsibility to safeguard dignified working conditions, in accordance with its institutional duty to represent the interests of its members.

Therefore, and with respect to companies, unions are expected to defend the rights of their members while considering the sustainability of the company. Unions, therefore, have an institutional responsibility not for the benefit of shareholders or owners, but for the importance of the company as a workplace that fosters employability and professional development. Some legislation, as a result, defines union activity as being oriented toward negotiating wages, working hours, access to education and professional training, occupational safety and health, and equal treatment—ensuring that all of this aligns with the economic and operational viability of the company.

In short—and as labor legislation indicates—, employers and unions can agree on fair wages and adequate working conditions as the basis for maintaining good labor relations, all while ensuring alignment with the company's need for sustainability. This, as we have said, is a matter of justice.

In this way, negotiation (*i.e.*, the process through which an agreement can be reached between two or more parties) depends not only on the relative advantages of the parties involved but also on the techniques used to gain advantages in the agreement. However, any agreement reached through negotiation, when considered ethically, is a matter of justice. In other words, an agreement resulting from a negotiation is ethically good because it is just. In this sense, abuse of power or dominant positions in a negotiation process is not inherently bad; they are wrong when they lead to unjust agreements or prevent the more advantaged party from addressing the needs of the other party, which should compel them to negotiate.

In business and economic activities, there are two types of agreements resulting from negotiations, namely, commercial agreements and collaboration agreements. The former are characteristic of market or exchange relationships, governed by the principle of mutual benefit and commutative justice, as previously discussed. The latter, however, are characteristic of employment relationships and are governed by the principle of merit, corresponding to contributive-distributive justice.

In this regard, any process of union negotiation should primarily be guided by the principle of contributive-distributive justice, as union members are in fact employees. In other words, union negotiations should aim to reach a fair collaboration agreement between workers and employers based on the principle of merit and the distribution of benefits according to employees' contributions. This also includes considerations of fairness in compensation. Thus, in accordance with these two principles of justice—merit and amount—the primary objective of collective or union negotiations should be the recognition of employee merit and, at a minimum, ensuring wages are aligned with market standards; as we have explained before for economic amount in a compensation.

On the other hand, and due to the nature of union activity in its guild role, negotiation processes also acquire a political or legal dimension concerning social justice. Indeed, the union's mission is not only about balancing the relationship between employee and employer but also about enhancing workers' well-being in both their professional and personal lives, that is, employees as citizens. Nevertheless, in matters of justice, not all aspects of a worker's well-being fall under the employer's responsibility. Therefore, it makes sense for unions to finance certain worker benefits outside the company's budget. In fact, in their guild dimension, unions often function as mutuals or cooperatives, that is, they provide benefits with resources coming from the workers themselves or, in other cases, public funding. The latest, however, is a matter of social justice in which employers have no significant role in what can be fairly demanded from them.

In short, an ethics of bargaining for unions can be framed under the criterion of justice, whether distributive-contributive or social, in terms of what benefits workers in their role as citizens.

5.3.2 *Practices Associated with Bargaining*

In light of the above, the topics related to the social dimension of work will focus on the role of unions. While other aspects of work—such as unemployment, technology, and occupational health—also play significant roles in the social consideration of employment, we will concentrate on workers' associations, as many of these other issues have already been addressed within the context of organizational ethics. Indeed, it can be argued that every aspect of personal work carries a social dimension. However, to highlight areas not previously explored, it is important to emphasize the social value of workers' associations and their related topics.

5.3.2.1 Union Dominant Positions

Union's Organizational dominant position: An imbalance in bargaining or union power to the detriment of companies themselves refers to situations in which, either due to legislation or business circumstances, unions have excessive bargaining power in relation to companies. Thus, companies can only follow union instructions, which in practice is not a negotiation, but rather subject to workers' commands.

Union's Industrial sector's dominant position: Similarly, a sectoral imbalance is produced when unions, grouped into general guilds or by industry, can negotiate with individual companies. Therefore, they have power over companies that are unbalanced and do not really allow for bargaining.

Union's Political dominant position: A political imbalance may arise with a partisan conflict of interest, hidden lobbying, or peddling of influences. This usually happens when union leadership is explicitly or implicitly linked to political parties or political authorities that are involved in labor legislation or collective bargaining processes.

The actions of unions not only have an impact on the lives of workers but also on the sustainability of the company, productive sectors, and society as a whole. In this sense, just as we do with corporate citizenship or the social impact of a company, it's possible to consider some criteria of union responsibility that should guide the decisions of workers' leadership. These forms of responsibility can indeed be considered based on the interests or stakeholders of the unions themselves.

Union's Civil Responsibility: Similarly, strikes and protests, which are legitimate tools, should not cause significant disruption to citizens' daily activities. This means, among other aspects, not blocking traffic, respect for private property, the protection of public property, etc.

Union's Institutional Responsibility: Trade unions, in the legitimate protection and extension of labor rights, cannot act in such a way that their demands pose a risk to the sustainability of the company. Union or guild leadership must ensure the continuity of companies since they are ultimately linked to labor and social development.

Union's Responsibility for Consumers: The legitimate right to strike, which is a fair negotiation tool, must be used in a way that is responsible toward consumers. When activities are suspended, companies must still be able to meet the basic and urgent needs of the population. For example, health care guilds, even when they cease their activities during negotiations, must maintain ethical shifts, given the importance of the service they provide.

Union's Responsibility for Employability: Trade union activity must aim to improve social rights, not just increase them. This means that it is important for changes in labor law to improve forms of work for the benefit of both companies and workers. However, when legislation is tightened excessively, this can entail an exaggerated increase in costs, impeding the continuity of companies or artificially increasing the recurrence of labor disputes. All this has a natural effect on a company's employability conditions and can affect not only the industry's hiring decisions but also formal employment.

Union's Political Responsibility: The risk of conflicts of interest on the part of union and guild leaders occurs when there are relations of interest between these same leaders and political parties. When legislation does not explicitly prohibit it, it must be avoided because it is an ethically illicit practice. Self-regulation of unions themselves is best, excluding themselves from partisan participation for the sake of the public good and national institutions.

5.3.2.2 Union's Political Conflict of Interests

Concerning political authority, unions maintain a relationship that more closely resembles that of a workers' guild than that of a union. This relationship entails representing workers' interests in the development, implementation, and oversight of labor legislation and public employment policies. Guilds, in this sense, are responsible for ensuring compliance with labor or social security laws, reporting violations to administrative or judicial authorities, and representing social interests compromised by the failure to enforce protection laws established for their members, either jointly or separately from state services.

However, this does not imply that union and guild leaders should have declared political affiliation. Membership in political parties can lead to significant conflict of interest, especially when union leaders share party affiliation with political authorities.

On a more concrete level, union activity can be analyzed in terms of limiting power imbalances and its impact on stakeholders. This helps us understand the just role of unions, aligned with the necessary responsibilities their leadership must assume toward both companies and society.

5.3.2.3 Union Busting

Finally, we must consider anti-union or union busting practices, i.e., those in which companies engage when they restrict and/or prohibit legitimate union activity. This requires us to consider what legitimate union practice is.

Union practices are legitimate when, based on the right to associate, they aim to achieve effective labor rights on balance with the interests of investors, managers, and consumers. Likewise, the legitimacy of union activity is achieved when labor rights are in harmony with the good of citizens, the public good, and the political good.

All managerial practices and policies that seek to diminish the effectiveness of unions in terms of labor rights must be considered union busting, which is illegitimate from an ethical point of view because it does not respect the human rights associated with labor activity and the rights that emanate from it.

Labor rights are all those natural principles that, derived from the principle of justice and the common good, apply to the world of labor relations in any circumstance or society.

For example, labor rights include the right to strike (as no one can be forced to work), the right to fair remuneration (as no one should be left without compensation or with unjust compensation), freedom of association, the right to rest (which comes from the right to participate in the common good), and the right to safe working condition (which comes from respect for physical and psychological health).

Union busting is not only illegitimate in regard to the law or the means—when it is crafty—but essentially insofar as it actively restricts the exercise of natural labor rights.

5.4 The Social Dimension of Markets

Following Aristotle, Aquinas considers men as social animals that naturally live in community (Aquinas, *De Regno*, I, 1, 2). As a result, the acquisition of their means of subsistence—from making food to trading—requires cooperation (Aquinas, *in Pol.*, I, 1, 12), since no man alone can achieve what is necessary for subsistence. This was precisely the origin of the social division of labor presented by Aristotle (*Pol.*, 1252b; NE, 1133a), that allows the commercial realm to go beyond the domestic realm (Finnis 2011a).

The so-called intermediate communities (Aquinas, *in Pol.*, I, 1, 12) rely on the social division of labor to achieve an increased satisfaction of non-daily needs (Aristotle, *Pol.*, 1252b; *NE*, 1133a). As Finnis (2011a) explains, Aristotle's idea refers to the natural condition in which specialization, technology, coordination production and marketing activities, the use of complex means of exchange (money), and a formal market allow for a better life, that is, beyond what each family on its own could aspire to, or, in other words, an economy outside of the domestic sphere (Scalzo et al. 2017).

Aristotle explains that, although the family is a form of society given that it facilitates cooperation towards the satisfaction of needs, we must resort to other forms of societies that facilitate the satisfaction of needs outside the family sphere. Thus, the emergence of intermediate communities—those between the family and the political community (called the state in the modern era)—refers in this case to the need to develop societies that achieve results on the basis of the so-called social division of labor (Aristotle, *EN*, 1133a), that is, in the commercial relationship of farmers, doctors, builders, etc. In this context, the market is a more perfect form of cooperation for the provision of the basics for good living.

This natural tendency to live in society is justified by the basic practical need to receive—and provide—cooperation from other people to achieve the resources needed for a good living. Society, therefore, is essentially a realm of cooperation, expressed in terms of a network of mutual services, and such a feature is natural because every human person is naturally aimed towards a good life. Without cooperation, human society would perish (Aquinas, *S.Th.*, II-II q. 66, a. 1. ad. 6).

The idea of the market as a kind of common good requires two initial considerations, namely an understanding of what the common good is and in what sense the market is a common good (Pinto-Garay et al. 2023). Indeed, as John Finnis explains, the common good is not an empty concept or, we might add, purely conceptual. The common good refers to certain conditions of people's lives that are qualified as qualitatively superior. Finnis himself explains that the common good refers to the following: (1) obtaining the material resources necessary for the development of people's activities as a whole, (2) the coordination of those actions necessary for the development of joint activities, (3) the possibility of personal development for each and every member of a community (institution, family, company, corporation, etc.) through joint action, and (4) the effective improvement of each of these members, understood as their integral development (Finnis 2011a).

5.4.1 *Common Good and Participation*

In light of Finnis' description of the later common good, we can conclude that the market not only has a common good of its own that is evidenced in a 'market community,' but also that the market itself promotes the common good of other communities, such as family, companies, society, and the state.

Thus, the possibility for the market to become a community—i.e., to provide a common good—is based on its capacity to (1) provide sufficient material conditions for a good living; (2) to establish effective cooperation conditions; (3) to open up spaces of participation for all its members; and (4) to safeguard the development of its members in terms of the acquisition of goods and services that eventually aid in integral development. Therefore, the market is a common good not only because of its capacity to provide more than the family can, but also because it acquires certain formal conditions, explicit or implicit, that allow, facilitate, and stimulate more complex relations of cooperation. Thus, regulating the use of money, commercial norms, trade cultures, the values of entrepreneurship, etc., all bring to life a commercial society with its own elements that add up to a more specific form of common good.

Given that the provision of resources for the satisfaction of needs and the conditions of cooperation are good for all people, especially insofar as they allow for integral human development, access is thus essential in the sense that capable people should not be left out of the market based on barriers that markets themselves may impose. In this moral sense, barriers or limitations to market access come in three different forms: (1) the constraints imposed on the market by non-commercial communities (such as state action that may impede entrepreneurship or companies' commercial development), and (2) inactivity on the part of market players (usually companies) that do not work to make new and better products available to certain consumers at affordable prices or in nearby markets.

Another form of limitation for market participation—and, consequently, for markets to become a common good for society—is the achievement and the preservation of dominant positions or monopolies, that is, the capacity to impede the participation of other competitive businesses.

Monopolies—from μόνος and πωλέιν, meaning 'one' and 'sale' respectively—or dominant positions are characterized by having profit and power in a way that prevents competitive market systems from existing, that is, limiting the unwasteful production and distribution of goods in larger quantities and less expensively (Finnis 2011a). In this sense, monopolies in all their forms—i.e., arbitrary exclusion of citizens from markets through oligopolies, monopsonies, or duopolies—are morally illicit because they restrain the possibility of other citizens to make a social contribution by means of their own commercial activities and, for consumers, to reach affordable products and services needed to improve their living standards.

On the contrary, social justice demands the stimulation of competition and free markets as the most proven way to increase commercial participation and, consequently, the capacity to improve personal living conditions. However, from a social perspective, markets can make a substantial contribution to the public good not just by sustaining unrestricted free will for commercial practices.

As Finnis (2011a) explains in relation to private property, commercial activities, mutatis mutandis, grant market participants the right of first use and enjoyment of their rents. Therefore, market decisions should be made within the framework of reasonable autonomy and, in accordance with the importance of rent, driven by productivity and care. However, beyond reasonable benefits for co-dependents and co-owners, rents should ultimately be used for the benefit of society. In fact, those who deliberate with intelligence, honesty, and care about their actions find good reasons to respect and promote the well-being not only of themselves, but also of their family, neighborhood, and economic associates. Commerce, in this sense, is no exception.

This idea was realized through the Social Market Economy—or Ordoliberal policies—which, as an alternative to both neoliberal and socialist models, facilitated the reconstruction and economic development of post-war Germany during the second half of the twentieth century. This approach offers a conceptual framework for rethinking the economy, market, and enterprise based on the principles of the Social Doctrine of the Church. The model was implemented by those who assumed economic authority in West Germany. The success of their economic policies, which fostered social development through market processes, earned them recognition not only for their accomplishments but also for the simplicity and effectiveness of this model that harmonized economic and social progress.

According to Müller-Armack (1978), a leading figure in Ordoliberalism, the concept of the Social Market Economy integrates personal deliberation with state economic policy, yet the latter must align with the nature of the market. This means that the 'natural' processes of the market are not directly controlled but guided. For example, inflation control policies do not fix prices but develop monetary policy through autonomous Central Banks, while policies aimed at reducing unemployment do not mandate hiring or prohibit layoffs but instead encourage worker hiring and employability through education.

5.4.2 *Practices Associated with Market and Competition*

Several of the following practices have often been mistakenly associated with commercial ethics. While it is true that some of them are related to commercial practices, they should be regarded primarily as social issues, as they hinder participation and, consequently, limit the role of markets in fostering social development and the common good.

5.4.2.1 Collusion and Cartels

Collusion occurs when parties—in general, managers of a company or organization—establish coordination formulas to deceive a third party (customers, authorities, competitors, etc.). This coordination, usually secret, involves fraud or the obtaining of an unfair advantage over a third party with whom they share commercial or competitive relations.

In addition, collusion may involve dishonest cooperation between competitors in the form of a sporadic or tacit agreement, the purpose of which is the manipulation of prices in favor of colluding people or companies.

If collusive practice becomes permanent, it is considered more like a cartel. Cartel collusion does not necessarily imply an explicit agreement, but often arises from implicit non-competition agreements, especially on the basis of low prices or discounts as a formula to prevent the entry of new players in the market.

Thus, collusion or cartel associations imply artificial fixing of prices or compensation (salaries or fees), as well as a practice contrary to the protection of third parties (trustees) when one of the parties has a fiduciary responsibility.

Cartels, in turn, can be classified into several types. Indeed, there are price cartels, that is, when two or more competing companies agree on the prices of their products. There are also territory cartels, or those that focus on market shares, patents and technology, and sales channels. There are even super cartels, as is the case with OPEC.

Now, cartels and collusive practices are ethically illicit because they limit, prevent or destroy the participation of commercial agents; hence, the authorities go after them in most developed economies, that is, economies that defend competition as a public good and a condition of social development.

It is important to note the fact that political authority is not enough; self-regulation plays an important role, especially when companies understand the importance of competing for the general good, not only for increasing their profitability and securing certain commercial positions.

5.4.2.2 Interlocking

Another form of collusive practice originates in the context of so-called interlocked board of directors. This refers to the simultaneous participation of directors in corporate governments of multiple companies or institutions. Now, the participation of independent directors on various boards always generates an interlocking relationship. This is important to keep in mind to clarify that not all interlocking is bad practice. However, interlocking can indeed be a condition for conflicts of interest, collusive practices, or outright corruption.

Indeed, directors who participate in companies that compete in the same market can lack the level of loyalty that they owe to shareholders. Interlocking may also favor collusion agreements or the creation of cartels between various companies.

Interlocking can also foment co-opting relationships, which refer to when a company or financial institution places a director in the corporate governance of a company to which it has lent large sums of money.

In this way, financial institution generates an irregular form of control in its favor. However, if participation on the board of a bank or other financial institution is agreed to when the creditor company is in financial trouble, this form of interlocking is called monitoring, and it is a perfectly acceptable way of contributing to the survival of the company. On the other hand, interlocking can give rise to certain forms of corruption and undue social influence. Regarding the former, it is important that company directors do not participate in other corporate governance processes of risk rating institutions or state institutions responsible for guaranteeing free competition.

Similarly, when a director participates in the corporate governance of non-profit institutions or media companies, she/he can have the ability to exert undue pressure to improve the prestige or cultural/political influence of the company where she/he is also director.

5.4.2.3 Speculation

Speculation is the practice of seeking financial gain, usually risky, by buying and selling assets whose prices will fluctuate. This occurs, for example, with money markets and futures trading, especially in commodity markets. These actors who benefit from speculation not only work (hence this type of speculation is not renting without working), but often stabilize prices, as is the case with these futures transactions.

Seen in this way, speculative practices have no major problem from an ethical point of view, unless the purchase or the offer itself is made with a view to changing the value of prices in one's own favor. In other words, there is an illicit form of speculation that takes the form of buying and selling assets with the aim of generating an impact on prices and, subsequently, profiting from them. Speculation in this sense is illegitimate, because it seeks individual profit through market manipulation, affecting other economic agents.

This not only damages the competitiveness of markets and increases their volatility but can also affect the ability of some groups of the population, especially the poorest, to purchase basic goods when prices rise. Similarly, speculation can be aimed at lowering the prices of various products, even below their production costs. This, naturally, affects both large and small producers, and often increases unemployment.

5.4.2.4 Private Bribery

These are formal or informal, explicit or implicit agreements entered into between private parties, allowing a producer to take advantage of a distribution channel to the detriment of the competition and other consumers.

This usually includes prizes and bonuses on sales or privileged positions in showcases and plazas. These 'awards' are given to employees or owners of marketing companies, or to specialists, as occurs at times between pharmaceutical companies and physicians.

5.4.2.5 Restricted Information

Information that is classified as restricted should also be protected. This kind of information may have a potential impact on the market, but is not directly associated with the company's commercial, strategic, or productive activity. For example, the dissemination of rumors or other aspects of relevant market players' private lives should be restricted.

5.4.2.6 Manipulation of Information

This malpractice occurs particularly, but not exclusively, in the financial industry. It is wrong because professionals across the board are responsible for the market's integrity. It is especially reflected in the use and treatment of information with a potential impact on the value of an instrument or product, causing damage among consumers or to competition. However, this practice pertains to confidential (non-public) information, rather than to restricted information, for example, information about the company's profits acquisitions, joint ventures and mergers; innovation and products or processes, new licenses or patents; current auditing processes; changes in management positions; legal disputes; bankruptcies; etc.

5.5 Environment, Care, and Property

The environment, when viewed from the perspective of social justice, presents a preliminary problem: nature cannot be subjected to justice because, among other reasons, it lacks the capacity to claim rights or bear moral duties and responsibilities. Indeed, justice applies only to relationships between people, that is, to the interactions between free and rational parties who can act with reciprocity (Aquinas, *S.Th.*, II-II, q. 25 a. 3; *S.Th.*, I-II, q. 28, a. 2). On the contrary, nonhuman animals cannot act with justice toward each other or any person, i.e., caring for a person *qua* person in a reciprocal manner.

However, while this argument is correct, it does not prevent us from assessing the value of nature in terms of social justice and ethics. In fact, from a moral standpoint, the environment not only has inherent value but is also a matter of the common good. In other words, non-rational living beings and their environment may be morally indifferent to us, but this does not absolve individuals or societies from the responsibility or duty to care for and preserve nature as a good in itself.

5.5.1 The Common Good of Conservation

Humanity's responsibility to the environment is arguably grounded in the fact that people—political animals—belong to the natural world just as any other living creature does. The human condition is not only rational but also animal; humans are, at least in part, mammals. Therefore, contrary to the views of many philosophers who argue that humanity is fully separate from nature (Plumwood 1991) or that humans should generally refrain from intervening in the environment (Garrett 1991) for similar reasons, human life is already integrated into nature through processes of production and consumption. Animals consume plants and other animals to survive (see Aquinas, *S.Th.*, II-II, q.64, a.1), and the non-living components of the environment, such as water and oxygen, must also be consumed to sustain plant and animal life. Similarly, humans must consume natural resources to survive. However, human consumption—which can be potentially unbounded—must be limited by a principle of responsibility, defined in this case as conservation, which is based on two key arguments: the intrinsic value of nature and the common good.

First and foremost, the responsibility to the environment is grounded not only in the inherent value of nature—a value that holds cruelty to animals as morally reprehensible—but also in the understanding that securing nature's intrinsic value requires humanity to view the environment not merely as a tool for human benefit, but as a fundamental condition for human flourishing. The environment provides essential conditions for humans to meet both the contemplative and physical needs of well-lived lives (McLaughlin 2012; Aquinas, *S.Th.* I-II, q.4, a.5). Therefore, the opposite of conservation is depletion, where nature is seen not as a vital condition for human life, but merely as a resource to be exploited and exhausted.

Moreover, the value of nature itself as an argument for preservation implies demands to care for the *diversity* of nature. The value of nature is not just in its single creatures but also its organic multiplicity of creatures and their relations (Aquinas, *S.Th.*, I, q.47, a.1), that is, as an ecosystem. In this sense, the complexity and perfection of the environment can be recognized in its intrinsic value and dignity which, in practical terms, demands care.

Indeed, the word 'care' derives from the Latin *cogitatus*, meaning understanding, attention, reflection, or thought. This suggests that caring for something—or someone—is based on its development and well-being. Caring for the environment, therefore, involves safeguarding the existence and proper functioning of nature, supporting its development as an ecosystem rich in species and biodiversity, and ensuring its sustainability over time by maintaining its natural cycles and processes of reproduction. In this sense, care is about protecting nature's state, growth, and development. As we've mentioned before, since human beings are part of the environment, caring for nature is ultimately a form of self-care.

On the other hand, conservation is a matter of the common good and, therefore, a form of justice. As McLaughlin (2012) explains, conservation involves caring for present and future communities that will need to interact with the environment in order to thrive physically and intellectually. This means integrating the environment as a vital resource and a basic condition for scientific advancement. The connection between environmental conservation and the common good is not only about material resources for human life but also about what nature offers for the subjective aspects of living well. For instance, a good human life depends on the acquisition of knowledge and the experience of beauty—both of which are deeply influenced by nature, which serves as a fundamental field for scientific discovery and aesthetic inspiration. In other words, conservation requires care for nature not only because it is a source of survival but also because it provides essential elements for fulfilling human life in more subjective ways. Therefore, we should view the environment as a source of both resources and goods.

Moreover, Finnis (2011b) sustains that a complete understanding of the common good demands a responsibility for the realization of human goods in a way in which human flourishing in future states of affairs is intrinsically attached to our present action. Thus, we have the responsibility to avoid any practices that could impact the environment in a way in which the yet-to-exit-communities will have limited or inexistent resources to flourish. Indeed, if we act without any consideration to the wellbeing of other people in our interaction with the environment, we commit an act of injustice. For this reason, human interaction with nature must be limited by an ecological interdependence pertaining to the common good (Finnis 2011a), that is, preserved for other communities to have the natural conditions to thrive. In this line, conservation demands society to safeguard the accessibility of other communities—present and future—to the environment as a condition for developing. In other words, this means the practice of facilitating—or at least not restraining—the possibility of other people and communities interacting with the ecosystem.

In fact, this is a major argument for moral grounding of sustainability. In the words of Finnis (2011a) social justice requires institutions and forms of life to satisfy the preconditions for long-term sustainable flourishing. In this way, resources are essentially a common stock for society—a material condition for achieving better standards of flourishing. However, their status as common stock does not imply they are of public dominion. On the contrary, the importance of natural resources for society's common good requires that they be primarily considered under the institution of private property. Ownership, in this context, carries a responsibility to the common good; without such responsibility, it is unjust (Finnis 2011a). Experience shows that the resources necessary for human life and well-being are best managed, developed, and made accessible when owned by specific administrators (Finnis 2011b). Therefore, property is the most effective way to manage natural and other resources needed to achieve higher standards of living and the common good. In other words, accessibility does not imply public ownership or the absence of property, nor does it reduce property to individual ownership alone. Accessibility, instead, provides an ethical justification for private property as a means of managing and making available what is necessary for societies to achieve better standards of the common good (Finnis 2011a).

Indeed, the administration of natural resources as property is morally good when it fosters the accessibility of the community to those goods through the assurance of order in the use and consumption of those same natural goods. Property, therefore, is conceived as morally licit only when is used to induce more and better general interaction and participation with those natural resources—making other present and future persons and communities to be objective and subjective benefitted from its private possession and administration (Aquinas, *SCG*, III, 112.13)—but in a way in which the normal conditions of the environment are preserved in its interdependence, diversity, beauty, multiplicity, etc. In this sense, Aquinas (*S.Th.*, II-II, q.64, a.1, ad3) would consider morally licit to have property over natural realms, animals, or plants because private administration provides better conditions for wider access to the natural resources needed to achieve better standards of common good.

In this manner, the environment, even when it is not properly a stakeholder, can be considered indirectly as a concern for those stakeholders responsible for safeguarding private property as a means for keeping the best standards of accessibility to the environment. According to Aristotle (1995: II, 1263a), those who are entitled to safeguard private ownership as a means for the common good are the legislators. Therefore, the accessibility to private owned natural environments is a matter of legal-political concern, according to Finnis, a requirement of social justice (Finnis 2011a).

From a social perspective, access to environmental goods is made possible by a public order that streamlines their use and availability. In this regard, the social institution of private property plays an essential role by providing greater guarantees for the effective management of natural resources and facilitating access to various environmental benefits through the social, commercial, economic, and political order. Take, for example, popular science magazines like *National Geographic*, which, through the marketing of their products, grant broad access to knowledge about nature. The company that owns the magazine enables an organized use of nature as a source of knowledge.

However, this does not imply any form of private property that restricts exclusive use of natural resources; rather, it refers to private property models that ensure good management for the optimal use of resources. Indeed, private property understood as an inalienable right that guarantees individual enjoyment is not the most suitable approach to fostering a balanced relationship between the environment and the common good. Strictly speaking, this type of inalienable property is based on exclusion, not on better management or the general good.

Conversely, private ownership of natural resources is justified because it leads to better management compared to public ownership. This is because, in practice, public goods do not belong to anyone, even when it is widely believed to belong to everyone. In other words, the general use of the environment—even in its best state of conservation—requires the protection of private property as a social institution.

This form of private property, as the foundation of the so-called Social Market Economy, provides a more complete understanding of the proper relationship between nature and the common good. It avoids the extremes of private property defined solely as an inalienable individual benefit, as well as the restrictive logic of environmentalism that limits any interaction with nature, even when the most vulnerable communities have urgent needs.

5.5.2 *Applied Theories of Conservation*

In more applied terms, the contemporary discussion on conservation has been grounded in at least two central concepts: sustainability and the circular economy (Caradonna 2017). Other environmental topics—such as ESG, the Triple Bottom Line, or the Doughnut Economy—are, in fact, practices associated with conservation.

Now, these two concepts, sustainability and circular economy, are not moral principles—like, for instance, justice, the common good or property, among others—given, among other reasons, that their original motivations were not explicitly ethical.

However, even though sustainability and the circular economy are neither principles nor practices, it is essential to consider their theoretical foundations as they have significantly contributed to the contemporary discussion on the environment-business relationship. In fact, without a clear understanding of their original motivations and conceptual developments, it would be almost impossible to agree on what we mean when we explain them or when we intend to assess firms and institutions that have integrate them as strategic goals.[1]

[1] Future developments in a neo-Aristotelian approach to business ethics could indeed explore the possibility of grounding concepts such as sustainability and the circular economy in the principles of justice and the common good. However, as such an endeavor lies beyond the scope of this book, we will limit ourselves to outlining briefly the essential elements of how sustainability and the circular economy should be understood, based on their original formulations and early conceptual developments.

5.5.2.1 The Theory of Sustainability

The concept of sustainability has its origin in the work of Hans Carl von Carlowitz, who used the term for the first time (*nachhaltigkeit*) in his 1713 treatise *Sylvicultura Oeconomica*. Von Carlowitz was the son of a logger (forester) and, after his studies in law and cameralist administration, entered the civil service. At the age of 32, he was appointed a mine administrator, and, in 1711, he became the director of Saxony's flourishing mining industry, which employed more than 10,000 people and focused especially on silver mining.

Von Carlowitz proposed the concept of sustainability given his concern for the nascent social and economic consequences appearing as a result of the scarcity of wood, a phenomenon that had its origin in the local industry's decades-long consumption of native local forests. For von Carlowitz, this problem derived especially from an industrial practice characterized by short-termism and the search for immediate profitability, that is, by an industrial culture that ignored the scarcity of productive resources.

Von Carlowitz, in effect, witnessed how the industry, once nearby forests had been consumed and in the face of the problem of scarcity, began to intervene in river systems to use them as transport channels for distant resources. In this way, scarcity and transportation costs led to a significant increase in the price of wood, hitting both the industry and the population in general, which depended on wood for subsistence. In the eighteenth century, Europe wood was, in effect, what fossil fuels were during the Industrial Revolution. Thus, forests provided resources for heating and cooking for European families. Increased prices brought obvious consequences for the population's quality of life. In addition, groups that dominated the wood market began to appear in favor of their own industries, without concern for the population's wood consumption. Thus, as von Carlowitz evidenced, the timber problem became not just an industry issue, but one for Saxon society in general.

Thus, von Carlowitz's description can be summed up as a phenomenon of social and economic crisis resulting from the total absence of sustainability principles. This crisis had wide repercussions in different areas, including raw material, producing commodities, legal loopholes and lack of authority, globalized production, linear extractive activity, non-renewable resources, supply shortage, and inflation.

This century's old discussion remains relevant. Indeed, the same nine conditions described by the Saxon cameralist may be conceptualized a bit differently today, but they are still very similar to the original version. Today we could call it a Model of the Sustainability Crisis.

Indeed, the issue of forests is a matter of raw materials. The mining and metallurgical industry that von Carlowitz himself managed is, for its part, a kind of economic institution understood as mere industrial systems characterized by the production of goods with little added value and highly dependent on extracted resources. Political authority, on the other hand, is nowadays understood as legal loopholes and a lack of authority. The issue of the rivers that appear later has a modern correlation in industrial relocation, that is, the search for productive resources and labor in other countries for reasons of cost. Then, and closely related to the latter, the use of new forests is a symptom of so-called extractivism, that is, the incorporation of natural resources in production systems without any consideration for the environment in a linear production model. Along the same lines, access to new forests is nothing more than the organization of business activity based on non-renewable resources. Finally, this entire approach to economic activity results in the scarcity of production resources (i.e., the problem of wood scarcity), which, in turn, translates into markets with very limited supply volumes, as well as price inflation that is the product of scarcity itself. All of the above results in a crisis that von Carlowitz describes in forestry, but that we know is common in other industries and societies.

Now, as we have seen, for von Carlowitz, a social crisis of sustainability is caused by a practice characterized by certain conditions described in 9 systemic, interrelated dimensions. Along these same lines, the creation of an ecosystem based on sustainability should be achieved by improving each of the nine elements that participate in said ecosystem.

Thus, the first basic element of an ecosystem based on sustainability proposes recategorizing raw materials as natural resources and the environment. This means that, although economic and social development must use natural resources, their use requires special care in terms of efficiency (using as little as possible and with a sense of preservation), as well as mitigation and environmental repair efforts.

Second, it is important to rethink the nature of economic institutions. This means advancing them qualitatively, leaving behind the idea that they are mere productive groups, and leading them towards business activity that defines their business model with less dependence on basic industrial processes and more oriented towards innovation and the generation of intangible added value.

From the perspective of society itself, legal protection, restrictions, regulations, but, above all, public policies are essential. This means that a good sustainability ecosystem not only integrates regulations that restrict or prohibit it but also generates initiatives to promote the development of a better, long term economy that is attentive to the efficient use of the environment.

Next, it is important to distinguish global production from global markets. The latter, in effect, is evidenced in the fact that a country not only contributes productively to other economies—be it labor, raw materials or other productive resources—but can also benefit from international trade relations, achieving better standards of wellbeing in the population through the consumption of value-added products and services. In this way, we can avoid situations in which poorer countries manufacture products or offer services from which they themselves cannot benefit, because they simply do not have access to them.

On the other hand, this approach proposes a circular economy instead of a linear production system. As mentioned, the former is oriented towards the reuse of raw materials and business models that incorporate product innovation as a commercial strategy.

Then, in terms of non-renewable assets, it is essential to rethink business models based on renewable productive resources. Faced with the certainty that many industries cannot achieve total independence from environmental resources, the alternative lies in the use of renewable resources and, furthermore, in the use of non-tangible resources as an important part of the business model. The latter translates, for example, into empowering human capital, knowledge management, etc.

Finally, all of the above basic elements of an ecosystem based on sustainability must bring certain visible results that are evidenced in the existence of balanced markets and robust purchasing power for economic agents. The former refers not just to the volume of products and services offered in a market, nor merely to the existence of a wide variety of products and services for both companies and consumers, but also to the stability of this offer. Simply put, by stability we mean the volume, variety, and availability of products and services.

Another visible and essential effect of a sustainable ecosystem refers to the ability of the various economic agents to participate in markets. The impoverishment of these agents causes precisely the inverse effect, i.e., the partial or almost complete marginalization of part of the population and their access to markets, which then negatively impacts personal wellbeing. In this way, purchasing power must not be solely effective based on the present needs related to living well, but must also gradually acquire certain guarantees of continuity over time, thus giving signs that society's quality of life is increasingly assured. When, on the contrary, facts and practices such as those described by von Carlowitz begin to emerge, a sustainability crisis is sure to follow.

Along these lines, the modern concept of sustainability, which reappeared in the 1980s with a lot of support from public policies and international development agendas, can be categorized as a systems theory through which societies understand the relationship between the environment, human wellbeing and economic systems (very much in the logic of Carlowitz's deforestation problems).

This systems theory—usually summarized with the three Es (Environmental, Equality and Economic), ESG (Environmental, Social and Governance) or three Ps (People, Planet and Profits)—is an interdisciplinary field of study that integrates ecology, economics, social sciences, human rights, engineering, history, ethics, and anthropology, among others.

Now, there are two large currents within the theory of sustainability. The first, defined as sustainability, refers more to conservationism, especially as represented by John Muir and the ecocentric or preservationist movement. The second, more accepted, approach is called sustainable development, where the principle of conservation is favorable to growth and economic development, although of course with a sense of social and environmental conservation.

The ideas that characterize the theory of sustainable development can be summarized as follows: Firstly, society, the environment, and the economy are necessarily interconnected. This is the idea of an ecosystem on which economic and social development is understood. Secondly, social and economic relations operate on the basis of environmental constraints. Without this consideration it is impossible to attain long-term achievements. Thirdly, the social good is achieved through the acquisition and preservation of long-term wellbeing, which has an intergenerational scope. Finally, the industrial society is sustainable when it is reformulated away from centralization and growth as a competitive advantage. The new industrialization path is more local, and growth comes through innovation rather than economies of scale.

5.5.2.2 Linear and Circular Economy

As Hanna Arendt (1958) explains, modernity is characterized by an expansion of consumption whereby products have progressively acquired a consumable nature. What had previously been built to last in the Industrial Revolution–and especially with the emergence of Taylorism and Fordism–is now taking on a more perishable nature through programmed obsolescence.

This change, however, is accompanied by the valuation of efficiency in production costs, that is, in economies of scale. Said economies are characterized by an increasingly positive relationship between mass use of productive resources, the production of increasingly cheaper units, and increased economic utility through the previously mentioned massification of consumption. Thus, the economy of the twentieth century is characterized, broadly speaking, by massification.

Adam Smith first outlined the advantages of this form of growth, but it truly spread in the twentieth century thanks to Alfred Marshall. It was precisely this economist who proposed the principle of marginal utility, not only as a mechanism that allows us to understand efficiency in terms of the cost associated with the production of an extra unit, but also as a way of identifying the indirect effects or positive or negative externalities associated with increased production volume.

This 'Marshallian' productive logic, which gained acceptance especially after the Second World War (1938–1945), is possible in the context of an idea of linear economic progress. The idea of linear progress first appeared in 1750 in Jacques Turgot's *A Philosophical Review of the Successive Advances of the Human Mind*. There, Turgot suggested a mechanistic model of progress in line with Newton's physics that coincided with the First Industrial Revolution and the global expansion of capitalism (*See* Turgot 1750/1973).

In the twentieth century, many began to see linear development as a reality of economic history because of the impact of unprecedented economic growth that had never been seen before in the history of mankind. The performance of European and non-Latin American economies set historical records and allowed for broad social development, as was the case for the baby boomer generation. In this way, the productive economy, as outlined by Smith and Marshall, allowed for the development of highly productive markets in the United States and Europe, especially during the second half of the twentieth century.

However, increased industrial production worldwide (especially with the contribution of Asian countries that, with their competitive labor costs, received Western industry) and economic growth based on intense consumption standards on the part of the global population also brought negative effects to society, especially when it comes to environmental matters.

These adverse effects led some to rethink the old Malthusian arguments regarding the economic and social impact of growth. An English clergyman and economist who was hired by the *East India Company* in London, Thomas Robert Malthus published *An Essay on the Principle of Population* (1872), where he argued that excessive (exponential) population growth is incompatible with a linear growth rate. Thus, he saw the idea of development that Turgot had prophesied as a logical contradiction because infinite growth cannot come from limited resources.

The 1979 oil crisis first brought this to the fore because it highlighted production's vital dependence on the availability of certain raw materials. This prompted some businessmen and academics to question the suitability of the linear model, especially in view of Malthus' criticism. This rethinking became a reality through the intellectual movement started with the Club of Rome and its publication of *The Limits of Economic Growt* (1972). This group aimed to spread the idea that the linear and infinite growth of production and prosperity was simply impossible in a world with finite resources (at least in the context of the technological development that society had achieved by the 1970s). Therefore, the capitalist economic model demanded reformulation.

Faced with this global problem, reformulating or improving the capitalist economic system without concession to the Marxist model became necessary. This, indeed, could be achieved within an understanding of the relationship between economy and society in which there is a place for both modern business institutions and corporations (as opposed to the Soviet Union's model, in which the state was the sole protagonist of the economy). For the Club of Rome, this reformulation of the model could be achieved in transition from a linear economy—e.g., neoclassical economy—to a circular capitalist economy (Colombo 2001).

This reconsideration, moreover, was meant to not only solve the problem of finite resources, but also other social and cultural problems that Western economies began to exhibit at the end of the twentieth and the beginning of the twenty-first centuries.

Indeed, the economic model that has characterized Western countries since the fall of the Berlin Wall has become paradoxical: On the one hand, it has significantly improved living conditions for the population, but on the other, it has become increasingly unsatisfactory, to the point of revealing a kind of social weariness with the market economy.

In very general terms, this is evidenced by the identification of economic (especially macroeconomic) progress with a consumerist society, increasing debt, social inequality, an overvaluation of the short term, and a view of the environment as mere raw material. All this was reason enough for the Club of Rome to push contemporary society to reformulate the capitalist economic model into a circular version in order to integrate normative elements beyond just increasing individual wealth.

Thus, the first problem to be addressed by a change of model is the relationship between production and the environment. To this end, the circular reformulation of the economy is not only defined by the substitution of raw materials for the reuse of obsolete products or waste but also introduces normative principles into society (or is based on them). These principles are associated with minimalism in lifestyle and consumption, institutional sustainability associated with extra-financial factors, and the relationship between business corporations and society in terms of permission and a social debt or mortgage.

Productive circularity—understood as the reintegration of old products into the productive process through the recycling of raw materials—makes sustaining the speed and intensity of consumption possible, avoiding waste production through reuse policies rather than by extending a product's useful life. In this way, productive enterprises stimulate the creation of new industries in the form of the reuse business, which is meant to progressively replace the raw material extraction industry.

It is also meant to impact consumer behavior, not only in terms of the growing preference for reusable, degradable products or products made directly from recycled materials, but also in terms of a new culture of use that does not throw away but gives away or (eventually) sells reusable materials. The criteria that prevail in this aspect of the circular economy is that of environmental care, which does not dismiss the idea of profitability outright, but integrates it into a new regulatory framework.

The environment, on the other hand, must be approached in two different ways: In terms of the instrumentality of resources (as an asset, if you will) or as a good in itself. The former is fairly easy to decipher and lines up with the first reformulation of natural resources from a linear economy to a circular economy: Natural resources must be considered a finite resource, and, moreover, environmental pollution produces extraordinary costs in productive processes. As for the limitation of resources, one strategy corresponds to technological development, which allows for a more limited use of resources with the same or better productive results.

An example of this is the development of the combustion engine which, thanks to technological progress, performs better and with fewer resources needed for its operation. Changes related to increased production costs caused by pollution must take place on the impetus of taxation or penalties that society implements to limit environmental damage.

But as a good in itself, the environment is not valued economically or as a production resource, but rather as a reality that manifests its value as a natural sight. For society, this also implies a reconsideration of the 'goodness of nature' as an area of knowledge, contemplation, and companionship, that is, as an essential factor in good living and to a certain degree of integral human development. Therefore, the climate and the environment are part of the common good because they dignify human life: We participate in nature, and it imprints a superior qualitative resource on personal, social, and cultural life that cannot be placed in a merely instrumental or productive context, even though it plays a fundamental role in the world economy.

The earliest and most procedural definition of circular economy likely came from the Ellen MacArthur Foundation. There, the model is shown as 'an industrial system that has a regenerative or restorative purpose and design.' In practice, this is conveyed in the search for and creation of opportunities for productive systems that do not incorporate obsolescence or lifetime limits, but rather a cradle-to-cradle system (Ellen MacArthur Foundation 2019).

This more detailed productive system depends essentially on the incorporation of renewable energies and the elimination of toxic resources and waste through the development of new materials, products, and systems. Therefore, the circular economy allows for a business model whose logic is oriented toward creating value through the recovery of the value retained in products that have already been used (Mentink 2014). It is an economic system based on closed circuits of production and use, with a harmonious relationship between ecology and society. In this system, biological resources necessary for industry can only be drawn from resources reintroduced into the biosphere without causing harm.

After the Club of Rome, this model was also developed as a paradigm for various schools of thought or theoretical models, including the following: regenerative design, performance economics, cradle to cradle, industrial ecology, biomimetism, natural capitalism, industrial metabolism, and industrial symbiosis. In addition, all these models have the following normative principles in common: design for reuse, the development of resilience through diversity, reliance on renewable energy, system design, waste as raw material, waterfall design, and shared values (symbiosis).

5.5.2.3 ESG and Non-Economic Performance Standards

Among various environmental standards, ESG has played a significant role, particularly in meeting the social demands that have prompted corporations to inform their constituencies with self-imposed, non-economic categories.

Within corporate strategies that combine a 'double bottom line' analysis of the twin corporate goals of profit maximization and social value creation (Puaschunder 2016), ESG (Environmental, Social, and Governance) goals have increasingly taken center stage in management literature and practice since Chris Yates published his ideas on responsible investment standards in 2013. Alongside other reporting standards such as GRI (Global Reporting Initiative) and the Sustainability Accounting Standards Board (SASB), ESG has incorporated additional principles to provide a broader understanding of the social dimensions of business practices, including ethics and sustainability (Armstrong 2020), innovation (Puzzonia 2018), competitive advantage (Bhandari et al. 2022), and even artificial intelligence (Burnaev et al. 2023).

However, while ESG has spurred several theoretical developments in management literature, it was not originally a theory but rather a common standard of practice derived from Socially Responsible Investment (SRI). According to Schueth (2003), SRI emerged in the 1960s, driven by the civil rights movements in the United States, which catalyzed the social demand for firms to consider factors beyond profits. Socially responsible investment—like many principles for social value creation—did not evolve from a pre-existing theory that subsequently influenced business practices, as seen with Porter and Kramer's Shared Value (2019) or Freeman's Stakeholder Theory (1984). Instead, it emerged as a movement initiated by society itself. As a result, ESG, like other social impact standards employed by corporations, lacks a solid conceptual framework at its origin. This gap has led to practical challenges, particularly in the area of corporate reporting, underlining the need for a more objective assessment of corporate social performance.

Corporate reporting refers to the periodic practice of informing a company's constituencies about self-imposed, non-economic standards and the achievements claimed under these standards. While this practice is expanding (for example, KPMG reports that the Corporate Sustainability Reporting Directive in alignment with ESG guidelines will be mandatory for nearly 50,000 European companies), these reports have often become sources of unreliable information, marked by inconsistent methods and little to no external verification (Yates-Smith 2013).

This issue arises partly because nearly every rating agency for social impact and sustainability uses distinct categories that span a wide range of projects, such as green initiatives, community services, responsible hiring, workforce practices, and reputation. However, these definitions often do not align with those used by other companies in the same industry. For example, the 'S' dimension of ESG encompasses various constituencies, including internal and external stakeholders, communities, consumers, and many others, but there are no explicit or general criteria for how these groups should be integrated or weighted.

Moreover, non-economic performance standards like ESG are often confused with policy (protocols, codes, or directives) developed through consensus, even when the theoretical basis for that consensus is at best unclear (Yates-Smith 2013). As a result, rather than representing a consensual policy, these standards may appear arbitrary. According to Crace and Gehman (2023), ESG performance reveals a confusing heterogeneity among firms, especially when influenced by a range of instrumental and normative criteria from diverse sources. In essence, when ESG (like other social impact standards) lacks an accepted theoretical framework, the ability to objectively inform empirical and qualitative performance standards becomes nearly impossible. Furthermore, the growing number of published corporate social reports, when lacking clear methodologies, becomes inadequate as a source for management and social science studies.

In this context, many authors have highlighted the challenges associated with corporate reporting on social impact, particularly issues related to subjectivity and inconsistency when comparing firms within the same industry.

References

Aquinas, T. 1274/1947. *Summa theologica* (fathers of the English Dominican Province, Trans.). New York: Benziger Brothers.

Arendt, H. 1958. The human condition. University of Chicago Press.

Aquinas, T. 1961. Liber de veritate catholicae Fidei contra errores infidelium seu Summa contra Gentiles. In *Sancti Thomae Aquinatis Doctoris Angelici opera omnia iussu impensaque Leonis XIII P. M. edita*, ed. P. Marc, C. Pera, and P. Caramello, 2–3. Torino: Marietti.

Aristotle. 1995. Nicomachean ethics. In *The complete works of Aristotle*, ed. J. Barnes, vol. 2. Princeton: Princeton University Press.

Aiyar, S. 2000. The problem of law's authority: John Finnis and Joseph Raz on legal obligation. *Law and Philosophy* 19:465.

Aquinas, T. 1272/2007. *Commentary on Aristotle's politics*. Indianapolis: Hackett Publishing.

Alzola, M. 2011. The reconciliation project: Separation and integration in business ethics research. *Journal of Business Ethics, 99*(1), 19–36.
Armstrong, A. 2020. Ethics and ESG. Australasian Accounting, Business and Finance Journal, 14(3), 6-17.
Bhandari, K. R., Ranta, M., & Salo, J. 2022. The resource-based view, stakeholder capitalism, ESG, and sustainable competitive advantage: The firm's embeddedness into ecology, society, and governance. Business Strategy and the Environment, 31(4), 1525-1537.
Burnaev, E., Mironov, E., Shpilman, A., Mironenko, M., & Katalevsky, D. 2023. Practical AI Cases for Solving ESG Challenges. Sustainability, 15(17), 12731.
Colombo, U. 2001. The club of Rome and sustainable development. *Futures* 33 (1): 7–11.
Caradonna, J. L., ed. 2017. *Routledge handbook of the history of sustainability*. London: Routledge.
Crace, L., & Gehman, J. 2023. What really explains ESG performance? Disentangling the asymmetrical drivers of the triple bottom line. Organization & Environment, 36(1), 150-178.
Dierksmeier, C., and A. Celano. 2012. Thomas Aquinas on justice as a global virtue in business. *Business Ethics Quarterly* 22 (2): 247–272.
Finnis, J. 1998. *Aquinas: Moral, political, and legal theory*. Oxford: Oxford University Press.
Finnis, J. 2011a. *Natural law and natural rights*. 2nd ed. Oxford: Oxford University Press.
Finnis, J. 2011b. *Human rights and common good: Collected essays*. Vol. 3. Oxford: Oxford University Press.
Garrett, J. 1991. Aristotle, ecology and politics: Theoria and Praxis for the Twenty-First Century. Social Philosophy Today, 6, 3-14.
Guillén, M. F. 2006. *The Taylorized beauty of the mechanical: Scientific management and the rise of modernist architecture*. Princeton: Princeton University Press.
González, A. M., & Scalzo, G. 2024. Circular subsidiarity: Humanizing work through relational goods. Business and society review, 129, 705-720.
Habermas, J. 1999. The European nation-state and the pressures of globalization. *New Left Review* 46:46–59.
Lovett, F. 2015. Civic virtue and political liberalism. *Political Theory* 43 (5): 619–642. https://doi.org/10.1177/0090591715603209.
Malthus, T. R. 1872. An essay on the principle of population or a view of its past and present effects on human happiness. In *An inquiry into our prospects respecting the future removal or mitigation of the evils which it occasions*, ed. T. R. Malthus. London: Reeves and Turner.
Malthus, T. R. 1872. An Essay on the Principle of Population Or a View of Its Past and Present Effects on Human Happiness, an Inquiry Into Our Prospects Respecting the Future Removal Or Mitigation of the Evils which it Occasions by Rev. TR Malthus. Reeves and Turner.
Müller-Armack, A. 1978. The social market economy as an economic and social order. *Review of Social Economy* 36 (3): 325–331.
Meredith, C. T. 2008. The ethical basis for taxation in the thought of Thomas Aquinas. *Journal of Markets & Morality* 11:1.
McLaughlin, R. P. 2012. Thomas Aquinas' eco-theological ethics of anthropocentric conservation. Horizons, 39(1), 69-97.
Mentink, B. A. S. (2014). Circular business model innovation: A process framework and a tool for business model innovation in a circular economy.
Macarthur, E. L. L. E. N., & Heading, H. E. A. D. I. N. G. 2019. How the circular economy tackles climate change. Ellen MacArthur Found, 1, 1-71.
Plumwood, V. 1991. Nature, self, and gender: Feminism, environmental philosophy, and the critique of rationalism. Hypatia, 6(1), 3-27.
Piketty, T. (2014). *Capital in the twenty-first century* (A. Goldhammer, Trans.). Harvard: Harvard University Press
Puaschunder, J. M. 2016. On the emergence, current state, and future perspectives of Socially Responsible Investment (SRI). Consilience, (16), 38-63.
Puzzonia, M. 2018. The Impact of ESG Investment, How Company and University can Collaborate to Realize It with Local Innovation. Journal of Intercultural Management, 10 (3), 171-194.

Pinto-Garay, J., Scalzo, G., & Schlag, M. 2023. Whose work? Which markets? Rethinking work and markets in light of virtue ethics. Business Ethics, the Environment & Responsibility, 32, 4-14.

Sombart, W. 1967. *Luxury and capitalism.* Ann Arbor: University of Michigan Press.

Schueth, S. 2003. Socially responsible investing in the United States. Journal of business ethics, 43(3), 189-194.

Sison, A. J. G., and J. Fontrodona. 2012. The common good of the firm in the Aristotelian-Thomistic tradition. *Business Ethics Quarterly* 22 (2): 211–246. https://doi.org/10.5840/beq201222215.

Sison, A. J. G., and J. Fontrodona. 2013. Participating in the common good of the firm. *Journal of Business Ethics* 113 (4): 611–625. https://doi.org/10.1007/s10551-013-1674-3.

Scalzo, G. 2017. A genealogy of the Gift. In Perspectives on Philosophy of Management and Business Ethics: Including a Special Section on Business and Human Rights (pp. 31-45). Cham: Springer International Publishing.

Scalzo, G., & Akrivou, K. 2020. Virtues, the common good, and business legitimacy. Jacob Dahl Rendtorff (Ed.) Handbook of Business Legitimacy: Responsibility, Ethics and Society, Springer Nature, pp. 1-13.

Scalzo, G., Pinto-Garay, J., & Akrivou, K. 2023. CSR and Virtue Ethics: The Common Good of Firms, Markets, and Civil Society. In The Routledge Companion to Corporate Social Responsibility (pp. 78-87). Routledge.

Turgot, A.R.J. 1750/1973. *A philosophical review of the successive advances of the human mind.* Cambridge: Cambridge University Press.

Yates-Smith, C. 2013. Socially responsible investment: good corporate citizenship or hidden portfolio risk?. Law and Financial Markets Review, 7(2), 112-117.

Wojtyla, K. (John Paul II) 1981. Laborem Exerc Esn: On Human Work. St. Paul Publications.

Chapter 6
Looking Ahead: Business Ethics and the Digital Technology Revolution

Abstract This concluding chapter reflects on the future of business ethics in light of the digital technology revolution. Drawing a parallel with earlier industrial revolutions, it highlights how digitalization, big data, artificial intelligence, and social networks are reshaping business, society, and culture in deeply interconnected ways. These transformations raise challenges that go beyond organizational, commercial, and corporate ethics, requiring an integrated approach. The chapter examines some of the most pressing issues—including privacy and data ownership, algorithmic bias and simulated human contact, labor substitution and employment shifts, and the influence of social media on public discourse—culminating in the foundational question of whether artificial intelligence can truly be considered "intelligent." It argues that technology, while transformative, does not determine society unilaterally but emerges in constant interaction with culture and human values. Business ethics, therefore, faces the challenge of articulating principles that ensure technological progress contributes to human flourishing, justice, and the common good.

Keywords Business ethics · Digital technology · Artificial intelligence · Big data · Privacy, Social networks · Employment · Common good

The relationship between business, technology, and society is most clearly exemplified in the First Industrial Revolution. This does not imply that such a relationship did not exist before; on the contrary, it has been ongoing, particularly since the emergence of modern corporations in the sixteenth century. However, the First Industrial Revolution stands out as paradigmatic due to its scale and profound social impact.

Specifically, the First Industrial Revolution refers to the technological, productive, and social changes that began in eighteenth-century Britain. The term was first popularized by economic historian Arnold Toynbee to describe the English economic transformation between 1760 and 1840.

In this context, the concept of 'revolution' is derived not so much from the technological changes themselves but from the specific location where this transformation took place. This is why the so-called Second Industrial Revolution is associated with

J. Pinto-Garay and G. Scalzo, *Justice and Corporate Excellence*, Ethical Economy 72,
https://doi.org/10.1007/978-3-032-12189-9_6

industrial developments in other parts of the world, such as the United States and Eastern Europe, during the nineteenth century.

That said, the technological aspects of an industrial revolution can be broken down into several distinct but interconnected changes. These include: (1) the introduction of new raw materials into production processes, (2) shifts in energy sources or the adoption of new energy resources, (3) the use of new machines and machinery in production and labor, (4) new organizational structures in work processes, (5) innovations in transportation and communication, and (6) the application of scientific discoveries in industry.

It is the simultaneous and synergistic emergence of these technological changes that drives shifts in the socioeconomic and cultural dimensions of society. This does not imply that society and culture are solely or primarily shaped by technology. Rather, it suggests that examining how technology, business, and society interact is best done by considering the technological and productive changes that occur. While technological advances certainly impact culture and society, the reverse is also true—society and culture significantly influence technological developments. Thus, technology, society, and culture interact systemically, rather than in a linear cause-and-effect manner initiated solely by technology. The decision to start by studying technological change is an epistemological one.

However, to claim that all social change stems from new technologies would be a mistake. One way to verify this is by noting that many technological and industrial innovations have had little to no significant impact on society or culture. Contrary to the assertions of Klaus Schwab and the World Economic Forum, it is unclear whether we are truly experiencing a Fourth Industrial Revolution. What is certain, however, is that the twenty-first century has seen important technological and social changes. Yet, there remain key aspects that must be examined before we can confirm whether this change is truly revolutionary.

To approach the issue of the relationship between business, society, and technology in the classical sense—by first considering technological changes—we can focus specifically on what is known as digital technology. This is the type of technology that has been notably absent from the industrial, social, and cultural transformations of earlier periods.

The etymological root of 'digital' is *digitus*, the Latin word for 'finger.' In the middle ages, *digitus* referred to numbers ranging from one to ten, corresponding to the number of human fingers and toes. By the twentieth century, it came to denote the digital numbers used to manage databases, in contrast to the analog use of data.

Thus, the digital society refers to the use of digitally managed and marketed productive resources, that is, systems based on numerical data. However, the ability for digital products to create industries is not solely the result of the development of digital products themselves, but also of the Internet (the World Wide Web), created by the English scientist Tim Berners-Lee between 1989 and 1990.

In terms of business activity, digitalization ranks alongside major industrial revolutions such as the advent of steam or electricity. These transformative factors not only contribute significantly to economic activity, but also reshape existing products and services, create new ones, and dramatically alter social interactions. Indeed, we now live in a digitally connected civilization with both economic and cultural implications. Digital production and marketing have redefined not just industry but also culture, civilization, and society, giving them clear social and business value.

In this context, such changes have given rise to important normative, regulatory, and legal challenges, particularly as they affect lifestyles, privacy, and personal and work relationships (González and Scalzo, 2024). Accordingly, the problem of a Digital Industrial Revolution cannot be framed only under commercial, organizational or public issues of justice, as we have done in this book, but as an integration of them in what we should define as a complex problem of business ethics as it integrates all organizational, commercial, and public stakeholders.

Indeed, following Sison et al. (2024) we could sustain that the Digital Industrial Revolution, and particularly the AI, presents important challenges listed as (1) foundational issues (in our approach, a matter of moral principles associated with the nature of human intelligence and its importance in matters of ethics), (2) privacy (a matter of public good), (3) bias and false human contact (particularly in commercial relationships and public administration), (4) employment (as in labor substitution and mechanization, i.e., organizational ethics), and (5) social media and public discourse (issues in public good).

6.1 Data and the Challenge of Privacy Protection

The term 'data' comes from the Latin *datum*, meaning 'a given thing' or 'an assumed fact.' Originally, it referred to 'facts based on calculation in mathematical problems.' By the nineteenth century, it came to signify 'numerical facts collected for future reference.'

During the twentieth century, the first computers were designed as tools for data processing and database management, involving the collection of numerically identified data on digital platforms. The exponential growth in data collection capacity, processing power, and storage technology has reached unprecedented levels. We now live in the 'petabyte era,' which is equivalent to 2 to the 50th power bytes. To put this in perspective, a byte consists of 8 bits, and the first computer developed by IBM in 1961 had a 64-bit capacity. The scale of growth is enormous.

This surge in capacity has created a need for sophisticated data analysis, a field generally referred to as data science. Data science involves not only managing data but also using it for decision-making, social studies, and predicting behaviors in areas such as markets and political elections.

Related ethical issues primarily concern the use of data that pertains to individuals' private lives. When considering the data life cycle, it's clear that each stage of data science interacts with privacy concerns. This creates an ethical dilemma around the ownership of personal data, especially as it pertains to privacy—defined as a domain separate from public or communal knowledge.

Privacy, from the Latin *privatus*, means something separate from what is public or common. It refers to aspects of life that are intimate and not meant to be shared or publicly known. On the other hand, anything made available to others, or common, refers to material goods and personal information that are accessible to the public.

What makes information or material good private or legitimately restricted from public use? The concept of privacy pertains to both the right of ownership over what can be used economically and to the intimate nature of one's life. Intimacy protects the space in which we express personal freedom, allowing us to live authentically without external pressures.

As Morente (1935) explains, private life must be understood in contrast to public life. Public life involves relationships governed by common roles or social functions, such as interactions with officials, judges, or merchants. Public relationships are impersonal, mechanical, and categorized. In contrast, private life is where we experience intimacy and express ourselves authentically, free from societal categories. Public life is defined by what is known about a person—such as their name, profession, or habits. However, knowing these aspects does not equate to knowing the 'who' of a person—the intimate, personal side of their existence. Public knowledge is thus about the 'what' (roles, characteristics, etc.), while private life reveals the 'who' (individual, intimate existence). Private life, therefore, is a space where individuals' authenticity thrives. Exposing this intimate space reduces or depersonalizes the person, turning their private existence into a public commodity. As García Morente writes, reducing a person to their public role or social function strips them of their individuality and authenticity.

Privacy also extends to lifestyle, as it is both a material manifestation of intimacy and an expression of how a person lives. Lifestyle incorporates both private and public aspects. The private lifestyle is a manifestation of intimacy, whereas public life is expressed in how one navigates social roles. Public life is tied to prestige, which is based on social roles and recognition.

When considering the ethical challenges of data use, it is crucial to distinguish between private life and public knowledge of one's personal information.

First, aspects of private life—such as lifestyle and intimate details—should not be made public without the individual's explicit consent. Even in public settings, only the relevant aspects of private life should be shared, such as when fulfilling legal or professional obligations. Second, activities within private spheres (home, marketplace, workplace, and civic life) should remain private and not be used by third parties without consent. Information from these spheres—such as purchases, personal habits, or family life—must not be exploited for purposes outside the intended public relationship. This principle extends to customer data in commercial settings, which cannot be used for purposes other than the intended transaction unless explicitly authorized by the individual.

In conclusion, while certain private data may need to be made public within specific contexts (e.g., for legal or professional reasons), it should be done with strict limitations. Personal data should only be disclosed to those directly involved in the specific public relationship and for the purpose at hand. Even in such cases, this data must not be used freely by third parties, ensuring that privacy is respected and protected.

6.2 Social Networks or the Impact of Digitalization on Sociability

All possible human luxuries were already present in antiquity, particularly in the Roman Empire—food, relaxation, decoration, and so on. It could also be argued that social networks are a completely new form of 'luxury' typical of post-modernity in the twentieth century. However, this perspective is mistaken, as the concept of social networks actually dates back to the Roman Empire, particularly through the importance of the roads built around the Mediterranean, which facilitated faster connectivity for public use. From Roman roads to the telephone, societies have developed various technologies to meet the need for connection. The major difference today is the digitization of connectivity.

Social networks, in essence, are platforms for connection between individuals, institutions, or both. From the beginning, with the rise of platforms like Facebook and LinkedIn, these social connection tools were designed around the principle of freedom of association and, later, the freedom of opinion.

Social platforms not only digitize existing social relationships, but they also transform and create new forms of interaction. This transformation involves not just the quantitative aspects of social interaction (such as the number of contacts or the potential for connectivity), but also qualitative changes in human relationships. Indeed, relationships of friendship, affection, intimacy, work, or business have been altered in both positive and negative ways.

Social networks have caused a kind of reformulation of pre-existing social structures. This refers to the gradual replacement of deeper, more integral human relationships with shorter, more numerous, and often superficial ones. This shift is driven by the convenience of digital connections, as well as the addictive nature of the graphic structure and informational architecture of social networks. Much like the rise of fast-food culture, we have now integrated 'fast friendship' into our lives.

The existence and development of social networks are driven not only by the convenience they offer for rapid social interactions—particularly the efficiency of social connection—but also by a business model that has become a key part of the global economy. This model relies on high volumes of digital relationships that are consumed and monetized, turning them into profitable business ventures.

Thus, the digitization of a fundamental human need—social connection—can be examined from the perspective of business ethics, specifically as a business activity. This examination can be approached by analyzing at list five aspects of the socio-economic phenomenon of social networks:

The first issue, in our view, concerns freedom of use. Access to social networks can be regarded as an indicator of social development. Services such as remote education and the expansion of online marketplaces offer considerable benefits to rural areas—assuming that internet connectivity and suitable devices are available. However, not everything about social networks is beneficial for users. While freedom is essential, safeguards for responsible use must be in place, including the restriction of certain content for specific populations, such as children. Freedom of use must be balanced with responsible access, which is not solely the responsibility of social networks or internet providers, but also involves local communities, family education, and personal accountability.

It is worth noting that increased access to social networks among children and adolescents has contributed to a rise in cyberbullying, with serious social consequences, including higher suicide rates. Thus, freedom of use does not automatically ensure responsible use, and it often results in harmful social outcomes.

The second issue relates to freedom to publish. Platforms such as LinkedIn, Facebook, Twitter, YouTube, and Instagram have been increasingly compelled to censor violent or false content. The social harm caused by such material—often disseminated anonymously—has led to the adoption of content restriction policies. However, these policies present further complications. One challenge is the absence of technology capable of autonomously filtering harmful content, which forces companies like Facebook to employ large numbers of human moderators. This, in turn, exposes these workers to disturbing material, causing significant psychological harm.

Another complication involves the spread of disinformation linked to political, religious, or social agendas. Social networks are intended to remain neutral and free from partisan influence, unlike traditional media such as television, print, or radio. Yet the rise of post-truth culture, fake news, and the lack of institutional support for fact-checking have compelled social networks to take positions—sometimes showing bias when censoring information that is not clearly false but unverifiable. This shift has turned social platforms into something akin to opinion sections, without the transparency of declared editorial stances. As a result, censorship inevitably implies a political or ideological position, which can undermine representativeness and balance, influencing electoral outcomes, the reputation of religious institutions, public health behavior, and more. This raises the question: can social media platforms truly maintain complete neutrality regarding control, censorship, and the delivery of information? Self-regulation is unlikely to be sufficient; some form of authoritative regulation may be necessary, though this could lead to the judicialization of digital relations, with significant social, economic, and political consequences.

Third, there is the problem of deterioration in social relationships. Greater connectivity does not necessarily lead to improved human relationships. Frequent interactions with family, friends, coworkers, or partners are not automatically enriched by social media use. For instance, "friendships" on platforms like Facebook often lack the depth of mutual knowledge, intimacy, and shared history that true friendships require. The same holds true for romantic relationships: while platforms such as Tinder may facilitate meeting potential partners, they often result in brief, casual encounters rather than stable, long-term bonds.

Although some couples do form meaningful relationships through these platforms, social networks generally appear to encourage transient connections at the expense of deeper ones. Indeed, many developed countries report a growing number of individuals living alone despite active participation in social networks. Likewise, Twitter has not substantially enhanced political participation, nor has LinkedIn significantly improved job quality. The prevalence of social networks may thus reflect, and even exacerbate, a broader decline in meaningful social connections, deepening social division and interpersonal alienation.

A fourth problem is digital addiction. Social networks and e-commerce platforms follow a design philosophy that prioritizes ease of use and intuitive navigation, known as user experience (UX). This approach not only ensures user-friendly interfaces but also leverages psychological stimuli to attract and retain users. The gratification of receiving likes, comments, or notifications can become addictive, particularly for individuals with anxiety or low self-esteem.

Addiction to social networks is a serious issue, especially among children and adolescents. Like any addiction, it restricts personal freedom and increasingly encroaches on more enriching aspects of life, limiting genuine engagement with the world beyond the digital sphere.

Finally, there are ethical issues concerning the business model of social network companies. Their model typically relies on paid memberships (e.g., LinkedIn Premium), advertising revenue from digital marketing, and the collection and sale of user data. This system thrives on strategic UX design that keeps users engaged, paired with user–machine interaction (UI) powered by algorithms tracking user behavior. These algorithms create detailed profiles of individual preferences, enabling highly personalized and targeted interactions.

In this vein, we need to consider that any data collected by these companies is in many cases not explicitly provided by users but is instead gathered through their interactions with the platform. This data is then sold to other companies, often for the purpose of enhancing digital marketing strategies. This practice can lead to the development of consumption habits that the user may not even be aware of, raising concerns about privacy and freedom, especially as it exploits patterns of behavior that were not voluntarily shared.

In sum, while social networks have brought about significant benefits in terms of connectivity and convenience, they also present a range of ethical challenges, from the erosion of genuine relationships to the exploitation of personal data for profit. Addressing these challenges requires a balance between freedom, responsibility, and regulation, ensuring that the digital realm remains a space for authentic connection and positive social impact.

6.3 Intelligence Can Never Be Artificial

So far, we have seen that the digital industry (Big Data and social networks) heavily relies on what we call artificial intelligence (AI) for its development.

The concept of AI is typically attributed to Alan Turing, John von Neumann, and Norbert Wiener. However, strictly speaking, AI first emerged in 1987 when Christopher Langton's article *Artificial Life* inspired a wave of research activities. Later on, AI research is highly heterogeneous, prompting the categorization of AI studies into three main branches: soft, hard, and wet.

Soft AI is mostly aimed at developing software by fusing not only in computer science, but integrating theoretical biology, cognitive science, and mathematics. In its more moderate scope of development, this AI is called 'weak,' and it aims to create digital simulations that exhibit properties characteristic of living beings, such as replication and autonomous improvement. In its strongest variant, Soft AI seeks to develop software as a replication of cellular automaton—formal systems capable of self-replication within an evolutionary framework, generating progressively more sophisticated systems with novel properties. One example of this technology is the dynamic storage capacity of a CPU.

Hard AI (short for hardware) refers to the development of robotics, that is, the production of agents capable of adaptive behavior, replication, and self-assembly, in accordance with principles of biological evolution. Significant advances have been achieved in areas such as molecular robotics and nanotechnology.

Wet AI concerns the creation of artificial chemical systems that exhibit defining traits of living organisms, particularly the ability to replicate and sustain themselves. These systems are often based on pre-existing organisms over which AI developments has focuses on the creation of the so-called artificial protocells.

From a different approach to said categorization of AI, this technology is defined in accordance with its capabilities, functionality and application.

In terms of capabilities, AI is classified as Narrow, General, and Superintelligent. Narrow AI or Narrow Artificial Intelligence (ANI) refers to systems that are designed to perform a specific task, like recognizing faces, playing chess, or generating text). ANI refers to AI systems designed for specific tasks—like facial recognition or recommendation engines—and is the current state of most AI applications today (Nilsson 2009). Generative AI belongs to this category because it specializes in creating content such as text, images, or music, but it cannot think or act beyond its specific domain. This AI can be seen in the services provided by companies like OpenAI, DeepMind, Anthropic, or Google DeepMind. General AI or Artificial General Intelligence (AGI) is supposed to have the flexible intelligence of a human being. These developments are offered also by OpenAI, DeepMind, Anthropic, among other companies. Finally, the so called Superintelligent AI or Artificial Superintelligent Intelligence, which would exceed human intelligence, is still theoretical, it does not exist (Russell and Norvig 2021).

In terms of functionality (i.e., how it operates) AI presents two existing types, and two, hypothetical. As for the first, AI operates as Reactive Machines or Limited Memory AI. Generative AI, for instance, fits into the Limited Memory category because it uses large amounts of data and context to generate its responses or outputs. Even though many companies and research institutions are trying new advances, Generative AI has not been able to understand emotions, intentions, or being conscious (Barredo Arrieta et al. 2020). The two hypothetical functions of AI are known as Theory of Mind AI and Self-Aware AI.

Finally, in terms of application, AI is described based on the technologies or applications it uses. Generative AI, for instance, relies on machine learning and especially deep learning, using complex models like neural networks. It often works through natural language processing (NLP), which allows it to understand

and generate human-like language. In image and video generation, it also overlaps with computer vision. Unlike expert systems, which rely on fixed rules, generative AI learns patterns from data. And unlike robotics, it typically operates in digital, non-physical environments (Goodfellow et al. 2016).

In this context, applying AI to companies raises at least three key ethical issues. The first one is a matter of responsible programming: When hard AI causes damage (e.g., a Tesla running over someone), the responsibility falls on the programmer and, by extension, the company. This argument is grounded in the idea that AI is not genuinely intelligent but merely calculating. Thus, the responsibility for any harm caused lies with those who programmed the AI system. Similarly, programmers and companies are accountable for the autonomous recommendations generated by soft AI, such as in marketplaces. For instance, if a system recommends adult products to minors, violates advertising laws, or presents misleading information, the programmer bears responsibility.

A second critical issue is a matter of replacing human decision-making by autonomous initiation of processes (Murcio et al. 2021). AI enables the extraction of organized knowledge from vast amounts of data, allowing for the identification of trends and patterns through complex probabilistic calculations. As a result, AI can initiate different autonomous processes of action—not as deciding, when machines cannot decide, but as original programed according to certain information inputs. This could happen, for example, when AI has the capacity to initiate preemptive attacks according to the way it has been originally programmed.

A third critical problem with AI is the progressive isolation from human contact when AI either simulates personal interaction in an imperceptible way or, in other cases, replaces human contact by simulating it with high realism. The first scenario raises issues of liability if, for instance, a bot provides incorrect or harmful information. It also prompts the question of whether users should be informed when they are interacting with a bot, given that human communication differs from machine interaction in many respects, not only in legal matters. While human conversations involve rational, free, and affective elements, machine communication is typically confined to delivering standardized information, regardless of its complexity.

In the second scenario, there is growing concern over the increasing number of people turning to generative AI for psychological counseling—particularly troubling when AI systems have, in some cases, advised users to commit suicide. In addition to this, it is important to consider what Sison et al. (2024) described as foundational issues, that is, the need to clarify if Ai is properly intelligent or, at least, if it has the capacity to achieve human standard of intelligence.

From our perspective, AI, in all its forms, is not strictly speaking a form of 'intelligence' but rather a material form of computation. While we might refer to it as 'intelligent' in the sense that computation is a part of human rationality, like imagination or memory, it is not accurate to equate AI with true intelligent thought. Imagination, calculation, and memory are integral to human thought, but they are not intelligence in all its potential.

Calculation relies on abstraction, allowing us to formalize concepts in logical or numerical terms. However, in the binary calculation that powered the first computers, a non-human or artificial form of reading emerged, capable of establishing numerical and logical relationships. These systems have advanced to the point where they can be programmed to improve computation and generate practical directions, such as in self-driving cars. While AI can adapt to complex realities with much faster computational capacity than humans, it does not truly understand what it is doing and, therefore, is not even human calculation.

In fact, human intelligence must be considered under what Finnis (2011), following Aquinas, describes as the human capacity for free, responsible, and intelligent decision:

> … What is to be done? What may be left undone? What is not to be done? We have, in the abstract, no reason to leave any of the basic goods out of account. But we do have good reason to choose commitments, projects, and actions, knowing that choice effectively rules out many alternative reasonable or possible commitments, projects, and actions. To have this choice between commitment to concentration upon one value (say, speculative truth) and commitment to others, and between one intelligent and reasonable project (say, understanding this book) and other eligible projects for giving definite shape to one's participation in one's selected value, and between one way of carrying out that project and other appropriate ways, is the primary respect in which we can call ourselves both free and responsible. For amongst the basic forms of good that we have no good reason to leave out of account is the good of practical reasonableness, which is participated in precisely by shaping one's participation in the other basic goods, by guiding one's commitments, one's selection of projects, and what one does in carrying them out. The principles that express the general ends of human life do not acquire what would nowadays be called a 'moral' force until they are brought to bear upon definite ranges of project, disposition, or action, or upon particular projects, dispositions, or actions. How they are thus to be brought to bear is the problem for practical reasonableness. 'Ethics', as classically conceived, is simply a recollectively and/or prospectively reflective expression of this problem and of the general lines of solutions which have been thought reasonable.

Anything besides such description, cannot be considered human and there is no possibility for any artificial machine to achieve this capacity to identify what is reasonable and intelligent to do in a personal project of life. In other words, machines cannot be ethical because they cannot be human.

References

Barredo Arrieta, A., F. Herrera, R. Chatilaf, R. Benjaminsh, D. Molina, S. Gil-Lopez, and N. Díaz-Rodríguez. 2020. Explainable artificial intelligence (XAI): Concepts, taxonomies, opportunities and challenges toward responsible AI. *Information Fusion* 58:82–115.

Finnis, J. 2011. *Natural law and natural rights*. 2nd ed. Oxford: Oxford University Press.

Goodfellow, I., Y. Bengio, A. Courville, and Y. Bengio. 2016. *Deep learning*. Vol. 1, No. 2. Cambridge: MIT press.

González, A. M., and Scalzo, G. 2024. Circular subsidiarity: Humanizing work through relational goods. *Business and society review*, 129:705–720.

Morente, M. G. 1935. Ensayo sobre la vida privada. *Revista de Occidente* 139:90–110.

Murcio, R., Scalzo, G., and Pinto, J. 2021. Can AI Emulate Soft Skills? The Future of Work in the Balance. *Soft Skills for Human Centered Management and Global Sustainability*.

Nilsson, N. J. 2009. *The quest for artificial intelligence*. Cambridge: Cambridge University Press.

Russell, S. J., and P. Norvig. 2021. *Artificial intelligence: A modern approach.* New York: Prentice Hall.

Sison, A. J. G., M. T. Daza, R. Gozalo-Brizuela, and E. C. Garrido-Merchán. 2024. ChatGPT: More than a "weapon of mass deception" ethical challenges and responses from the human-centered artificial intelligence (HCAI) perspective. *International Journal of Human–Computer Interaction* 40 (17): 4853–4872.

MIX
Papier aus verantwortungsvollen Quellen
Paper from responsible sources
FSC® C105338

If you have any concerns about our products,
you can contact us on
ProductSafety@springernature.com

In case Publisher is established outside the EU,
the EU authorized representative is:
Springer Nature Customer Service Center GmbH
Europaplatz 3, 69115 Heidelberg, Germany

Printed by Libri Plureos GmbH
in Hamburg, Germany